T0268263

CONNECTICUT
INVENTORS
AND
INNOVATORS

PETER HUBBARD

THE
History
PRESS

Published by The History Press
Charleston, SC
www.historypress.com

Copyright © 2022 by Peter Hubbard
All rights reserved

First published 2022

Manufactured in the United States

ISBN 9781467152099

Library of Congress Control Number: 2022943519

Notice: The information in this book is true and complete to the best of our knowledge. It is offered without guarantee on the part of the author or The History Press. The author and The History Press disclaim all liability in connection with the use of this book.

All rights reserved. No part of this book may be reproduced or transmitted in any form whatsoever without prior written permission from the publisher except in the case of brief quotations embodied in critical articles and reviews.

CONTENTS

CONTENTS

ACKNOWLEDGEMENTS

I would like to thank the following people for their help on this book: David N. Mullany for furnishing photographs and information on WIFFLE® balls and David A. Mullany for reviewing the manuscript; Bill Piedra for information on his invention; Alex Dubois and Sean Kunic of the Litchfield Historical Society; Bigelow's director of marketing, Mary Lauren Factora, for furnishing photographs and information; Tom Hennessy of the Lock Museum of America; Tom Borysiewicz and Kristin Silvia of the Bethel Public Library; Patti Philippon of the American Clock and Watch Museum; and Tessa O'Sullivan of the Hartford Medical Society. I especially appreciate History Press editor Michael G. Kinsella's invaluable support and guidance on this project. I also thank editor Hilary Parrish for her outstanding work on this book.

Most of all, I would like to thank my father and mother for their extensive contributions.

INTRODUCTION

I do not think there is any thrill that can go through the human heart like that felt by the inventor as he sees some creation of the brain unfolding to success.
—*Nikola Tesla*

In 1898, the United States Patent Office released an annual report that stated that in proportion to population, more patents were issued to citizens of Connecticut than to those of any other state. (At the time, the United States was composed of forty-five states.) In 1897, 1 in every 786 people in Connecticut was awarded a patent. Number two in issued patents was Massachusetts, followed by Washington, D.C., New Jersey, Rhode Island and New York.

But the 1800s was not Connecticut's only period of great innovation. In 2020, Bloomberg's annual State Innovation Index ranked Connecticut as the fourth most innovative state. To reach that mark, the state ranked in Bloomberg's top ten for patent activity, productivity and research and development intensity. Only California ranked higher than Connecticut in patents awarded per one thousand occupations in science and engineering.

With such a large selection, a determination needed to be made on which inventors and innovators to include in this book. Some were born, raised, created their invention(s) or innovation(s) while in Connecticut, lived their entire lives in the state and died and were buried in Connecticut. Others, such as Lewis Latimer, lived and worked for only a few years in the state, but those years were some of his most productive. On the other hand, some were born in Connecticut and left at an early age to carry on their education, life and work elsewhere. Lastly, there are some who only received their college or university degrees in the state and then left. Most notable in this category are the many scientists, researchers, medical doctors and so on who attended Yale University as students. The decision was to make mention of most noteworthy people but to primarily concentrate on those who had the longest and/or most significant connection with Connecticut.

Interestingly, there isn't a widely accepted term for someone who is from Connecticut—whether born there or a past or current resident. One term that is frequently heard is "Nutmegger" because Connecticut's unofficial nickname is the Nutmeg State. This is thought to be derived from eighteenth-century Connecticut peddlers who sold nutmeg house to house. The 2020 edition of the U.S. Government Publishing Office Style Manual specifies how the federal government will officially refer to natives of each U.S. state. For Connecticut, it chooses the term "Connecticuter." In the current volume, we will too.

U.S. Patents

Many of the inventions mentioned in this book were awarded U.S. patents. However, many, for various reasons, were not. To understand what a patent is, the best source is the United States Patent and Trademark Office. Its official website specifies:

> A patent for an invention is the grant of a property right to the inventor, issued by the United States Patent and Trademark Office. Generally, the term of a new patent is 20 years from the date on which the application for the patent was filed in the United States or, in special cases, from the date an earlier related application was filed, subject to the payment of maintenance fees. U.S. patent grants are effective only within the United States, U.S. territories, and U.S. possessions. Under certain circumstances, patent term extensions or adjustments may be available.
>
> The right conferred by the patent grant is, in the language of the statute and of the grant itself, "the right to exclude others from making, using, offering for sale, or selling" the invention in the United States or "importing" the invention into the United States. What is granted is not the right to make, use, offer for sale, sell or import, but the right to exclude others from making, using, offering for sale, selling or importing the invention. Once a patent is issued, the patentee must enforce the patent without aid of the USPTO.

Chapter 1

CONSUMER INVENTIONS

C onnecticut's residents have always been at the forefront of trying out the latest inventions. The state's location between the metropolitan centers of New York City and Boston has meant that its people have had relatively quick and easy access to the latest news and a wide selection of merchandise. The state's numerous institutions of higher education have also helped foster a rich environment for the inventor and innovator.

CLOCKS

Eli Terry

Often called the father of the United States mass-production clock industry, Eli Terry was born in East Windsor, Connecticut, three years before the beginning of the American Revolutionary War. At age fourteen, he was apprenticed to a master clockmaker and worked on both brass and wooden movement clocks. Seven years later, he opened his own business in Plymouth, Connecticut, a town north of Waterbury. He used wooden gears in his clocks because wood was readily available and inexpensive. At the time, according to clockmaker Chauncey Jerome, clock "wheels and teeth had been cut out by hand; first marked out with square and compasses, and then sawed with a fine saw, a very slow and tedious process." Clockmakers would custom make each clock and then deliver them a few at a time to customers.

This Eli Terry water wheel in Plymouth, Connecticut, once supplied power to the Terry Clock Shop. *Author's collection.*

In 1797, the United States Patent Office granted twenty-five-year-old Eli Terry a patent. It was the first U.S. patent for a clock. Terry would make a few trips each year to sell his clocks at about twenty-five dollars apiece. Chauncey Jerome became an employee of Terry's clockmaking in his early twenties to make cases for Terry's clocks until he decided to start his own case-making company. In Jerome's 1860 book, *History of the American Clock Business for the Past Sixty Years*, he related this story that was told to him by an elderly man named Blakeslee:

> *One day Mr. Terry came to the house where he lived to sell a clock. The man with whom young Blakeslee lived, left him to plow in the field and went to the house to make a bargain for it, which he did, paying Mr. Terry in salt pork, a part of which he carried home in his saddle-bags where he had carried the clock.* [Terry] *was at that time very poor, but twenty-five years after was worth $200,000, all of which he made in the clock business.*

In the first few years of the 1800s, Terry made three major innovations: using water power to run his factory, creating standard parts, and using machinery to manufacture his clocks. Most of his clock mechanisms could fit inside a box, making them perfect for placement on shelves or mantels. In 1807, Terry entered into a contract to make four thousand clocks—and then proceeded to fulfill the agreement. Three years later, Terry sold his business to two of his workers, Seth Thomas (who would go on to found the Seth Thomas Clock Company in 1853) and Silas Hoadley.

Although he designed and built many types of clocks, Terry's specialty was the one-day wooden shelf clock. Between 1816 and 1826, he received seven U.S. patents for thirty-hour wooden clocks. (These were one-day clocks with a few extra hours to allow time for their owner to wind them for the next day.)

Ultimately, his clock sales made Terry a famous and wealthy man. His innovations made it possible to sell quality clocks at lower prices, which, in turn, changed clocks from being a luxury item that only the wealthy could afford to a common item in the average home.

One of Eli Terry's original clocks graces the front of the First Congregational Church of Plymouth. *Author's collection.*

Terry's financial success inspired scores of other Connecticut entrepreneurs to become clockmakers. The nearby town of Bristol became known as the clock capital of the world. Terry's success with the interchangeability of parts was copied nationwide for a wide variety of products.

Terry passed away at age seventy-nine in 1852 in Plymouth. Today, one of Terry's original clocks—a Roman-numeral clock dial—sits above the entrance to the First Congregational Church of Plymouth. Its clockwork was made by Eli Terry late in his life. Plymouth's village of Terryville was named after Terry's son Eli Terry Jr. Today, the American Clock and Watch Museum in Bristol has an interesting collection of many Terry clocks, including tall clocks, grandfather clocks and shelf clocks.

Vacuum Cleaner

Ira Hobart Spencer

Inventor Ira Hobart Spencer founded the Spencer Turbine Company in Hartford, Connecticut, in 1892. In 1905, he introduced a turbine vacuum cleaner and founded the Spencer Turbine Cleaner Company.

Born in Barkhamsted, Connecticut, in 1873, Spencer attended school in Winsted for a short while and then, at age thirteen, moved to Hartford. As a janitor and organist at St. James Church in Hartford, he invented a water-powered hydraulic engine to pump air. It would replace the manually pumped pipe organ at his church. The demand by other houses of worship was substantial. When electricity was introduced, Spencer used electric motors. Spencer's engines were named Orgoblos, a combination of the words *organ* and *blowers*. Today, according to the Spencer Turbine Company's website, they still sell spare parts for some of the early Orgoblos.

By 1909, Spencer had forty-two patents awarded or pending on his vacuum cleaning devices and another twenty on his other inventions. The stationary vacuum cleaner systems developed by Spencer were also incredibly popular in the United States. Two of his large centrifugal blowers were installed at the Mormon Tabernacle Organ in 1915, and its central vacuum cleaning system was installed in the White House in 1925. The system was installed in New York City's Chrysler Building (at the time, the world's tallest building) in 1930 and, in 1931, in that building's world record successor, the Empire State Building. Many years later, a Spencer system was used in the Hartford Transit Authority's buses.

When he died in 1928, Spencer had over one hundred U.S. patents. He was buried in West Hartford. In 1975, the Spencer Turbine Company opened a new headquarters building in Windsor, Connecticut. It's still there nearly half a century later.

Ironing Board

Sarah Boone

Born into slavery in North Carolina in 1832, Sarah Boone, when about twenty-four years old, made her way to New Haven with her thirty-six-year-old husband, James (1824–1876), and her sixty-three-year-old mother, Sarah Marshall (1793–1868). Once in Connecticut, she set about obtaining the education she had been deprived of while enslaved. For decades, she worked as a dressmaker. With that experience, combined with her own creativity, in 1892 at age sixty she was awarded a patent for her invention of an early ironing board. It's said that a number of its features—including the padded exterior and the collapsible design—directly led to today's ironing boards.

In her patent application, Boone states that her objective was to "produce a cheap, simple, convenient, and highly effective device, particularly adapted to be used in ironing the sleeves and bodies of ladies' garments." Her ironing board was narrow with curved edges shaped like the inside and outside seams of sleeves. It was a great improvement over the simple wooden boards used by dressmakers and homemakers of the day.

Sarah Boone and her husband had eight children and lived in a home they owned on New Haven's Winter Street. As a member of the Dixwell Avenue Congregational Church, Sarah Boone died in New Haven in 1904 and was buried in New Haven's Evergreen Cemetery next to her husband and mother.

Sewing Machine

Elias Howe

Machinist Elias Howe invented the first practical sewing machine. Unlike other attempts at creating a sewing machine, Howe used a lockstitch. This involved a needle with a thread going up and down and picking up a thread from the bobbin on the other side of the object being sewed.

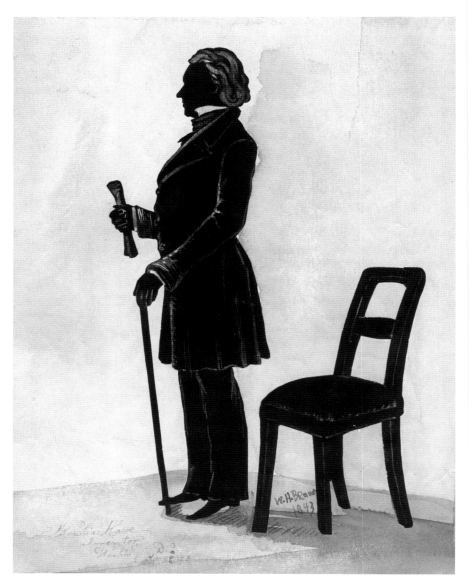

This silhouette of sewing machine inventor Elias Howe was created in 1843 by artist William Henry Brown. *National Portrait Gallery, Smithsonian Institution.*

Elias Howe probably built his machine in his spare time while living at a hotel at the corner of Route 44 and Bridge Street in New Hartford, Connecticut. In 1846, he was awarded a patent for his invention, but after he returned from an extended stay in Europe, he discovered other manufacturers were infringing on his patent. After a difficult legal battle, Howe won and secured his rights to the invention in 1854. From then until the expiration of his patent in 1867, he received royalties for every one of his sewing machines that was manufactured in the United States. The Howe sewing machine proved to provide incredible benefits both for manufacturers of clothing and for the improvement of sewing in the home.

At the beginning of the American Civil War, then forty-three-year-old Howe enlisted as a private in the Seventeenth Regiment of the Connecticut Volunteer Infantry and used his own funds to purchase most of its equipment. Poor health prevented him from seeing action.

After the war's end, Howe opened a factory along the Pequonnock River in Bridgeport, Connecticut, and manufactured as many as four hundred sewing machines per day. Eventually, he became one of the wealthiest people in the United States as he received five dollars for every machine sold domestically and one dollar for every machine sold in foreign countries. Today, a large statue of Howe stands in Bridgeport's Seaside Park. Perhaps the most unusual modern footnote to Howe's story occurred a century after he established his Connecticut factory. In 1965, the most famous British musical group, the Beatles, released its musical comedy movie *Help!* At the end of the film, the following line appears: "This film is respectfully dedicated to the memory of Mr. Elias Howe, who, in 1846, invented the sewing machine."

ICE-MAKING MACHINES

Alexander Catlin Twining

Inventor Alexander Twining's best-known inventions are in the area of ice-making and refrigeration. Born in New Haven, Connecticut, in 1801, he attended Hopkins Grammar School and earned a master's degree at Yale College in 1820. After graduating, he worked as a tutor at Yale for a few years. He then studied civil engineering at West Point in New York and taught and practiced civil engineering in Vermont. Twining then returned to New Haven in his late forties and stayed there until he passed away at age eighty-three.

Twining received a patent for his ice-making machine in 1853, which began the era of modern commercial refrigeration. He received additional patents for refrigeration and ice making in 1861, 1862, 1871 and 1874.

Although the Civil War prevented Twining from introducing his ice manufacturing equipment to the southern states, his methods prevailed and led to major change in the way ice was created.

Twining's interests were many, including the study of theology (he was a deacon at First Church on the New Haven Green for the last twenty-eight years of his life), astronomy and transportation (he was involved in the development of virtually every railway line extending from New Haven). Some of the other patents that Twining received were for a timber joining method, armor cladding for "resisting the impact of moving bodies and missiles," an improved caster and a railway signal and alarm that prevented accidents from open drawbridges.

ROTARY CAN OPENER

Ezra J. Warner and William W. Lyman

The first metal cans used for food storage were invented in 1810, but it wasn't until 1858 that Waterbury's Ezra J. Warner received a patent for his invention of the first U.S. can opener. The introduction of thinner metal cans helped make his invention feasible. Warner's opener consisted of a metal piece that looked like a bent bayonet that pierced the lid and a blade that sawed a line around the edge of the can. However, Warner's opener left a jagged edge and was so difficult to use that most storekeepers cut open the cans for the customers before they left their stores. The Warner opener was also used by the U.S. Army during the American Civil War.

The first opener that was like today's manual and electric openers wasn't patented until twelve years after Warner's device. Its inventor, William W. Lyman, was born in 1821 in Middlefield, Connecticut. A member of the locally famous Lyman farming family, he worked for many years as a pewtersmith in nearby Meriden. His invention, for which he received a patent in 1870, was the first rotary can opener with a cutting wheel. A safer instrument than Warner's, it was widely adopted for home use.

Prior to his can opener invention, lifelong Connecticut resident Lyman had been issued patents for a refrigerating pitcher (1858), an improved fruit can (1862), improvements in the manufacture of tea and coffee pots (1867) and an improved butter dish (1868). Lyman died in Meriden in 1891.

The major differences between most twenty-first-century openers and Lyman's invention were the later addition of a hand crank and a serrated rotation wheel, which held the edge of the can. They were added by San Francisco's Star Can Company in 1925.

MODERN TAPE MEASURE

William H. Bangs Jr. and Alvin J. Fellows

James Chesterman (1795–1867) of Sheffield, England, who made flat wire, put measuring marks on pieces of it and sold them to surveyors as Steel Band Measuring Chains. In 1829, he received a British patent for it. However, at about seventeen dollars each, the measuring tapes were too expensive for Americans, who were only paying about fifteen dollars for a bed with a bureau.

In 1864, Killingworth-born William H. Bangs Jr. (1835–1871) of the western part of Meriden, Connecticut, received a patent for a spring return pocket tape measure, which had the ability to lock the tape in place. In 1868, New Haven's Alvin J. Fellows (1841–1919) was awarded a patent for a "spring-click tape measure," which added a case, cover plate, click spring, lever and knob.

In 1922, Hiram A. Farrand received a patent for a spring pocket tape measure with concave-convex tape. Eventually, it was sold to New Britain, Connecticut's Stanley Works tool company. Today, most pocket tape measures use that design.

It took about twenty years after Farrand for the spring pocket tape measure to become more popular than the traditional folding wooden ruler. Today, even though digital (laser) measuring devices are growing rapidly in popularity, the tape measure designed by Bangs, Fellows and Farrand is a staple of every hardware store.

LIGHTNING ARRESTER

Alexander Jay Wurts

Born in 1862, Alexander Jay Wurts graduated from New Haven's James Hillhouse High School and received his PhD degree from Yale University in 1882. At Hillhouse, Wurts, along with three other students, founded high school–based Gamma Delta Psi Fraternity, which in future years would include as members Presidents Theodore Roosevelt and William Howard Taft, as well as General Douglas MacArthur.

After continuing his education at Stevens Institute of Technology and studying electrical engineering in Germany, Wurts went to work for the Westinghouse Electric and Manufacturing Company in 1887 as an engineer. It was there that he worked with George Westinghouse on the development of the Nernst lamp. Through Westinghouse, he met industrialist Andrew Carnegie. In 1904, Wurst was appointed the first faculty member of the Carnegie Technical School (later renamed the Carnegie Institute of Technology and today the Carnegie Mellon University College of Engineering). A professor of applied electricity, he headed up the school's electrical engineering department.

Probably Wurts's greatest claim to fame was his work on the Westinghouse lightning arrester, which diverts electricity surges—like those caused by lightning—and protects equipment. In 1890, Wurts was awarded the first U.S. patent for Westinghouse Electric for the lightning protection of power systems; two years later, Thomas Edison was awarded the first patent for an arrester for General Electric. For his work on "Lightning arresters for lighting and power circuits and non-arching metal," Wurts received the Franklin Institute's John Scott Medal in 1894.

REEL LAWN MOWER

Amariah M. Hills

About four years after receiving his first U.S. patent—for an improved lamp shade—forty-seven-year-old Amariah M. Hills of Hockanum, Connecticut, which is part of East Hartford, received the first U.S. patent for a reel lawnmower that mowed grass by hand.

After an apprenticeship in Hartford, at about age twenty-one Hills moved to New York, where he worked as a silversmith and watchcase maker for many years. In 1861, he began working on an improved lawnmower. Unlike some larger mowers (which were often pulled by horses), it was designed to cut small areas, such as residential lawns, by hand. Shortly before the patent was granted on January 28, 1868, Hills had manufactured almost five hundred of his mowers at the Curtis silverplate factory in Glastonbury.

In 1871, the Hills Archimedean Lawn Mower Co. of Hartford was established. Apparently, Amariah Hills had sold that company his patent

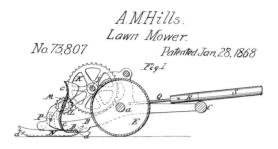

The 1868 patent drawing of Amariah Hills's lawn mower. Hills was born in a village of East Hartford, Connecticut, in 1820. *United States Patent and Trademark Office.*

rights around that time. The new company's Charter Oak model was released in the 1870s, and in the mower's first year, they sold almost $100,000 worth of them. The company went out of business in 1892.

STEAM COOKER

Alice M. Hobson

One of the early women inventors of Connecticut was Alice M. Hobson of New Britain. In 1891, she received a patent for a steam cooker. According to her patent application, her cooker provides a "simple and cheap article which can be quickly arranged for steaming several kinds of food at the same time, and which is provided with simple means, whereby the food in either section may be tested and its condition observed without disturbing the food in the remaining sections, and also to provide means for venting the kettle and collecting the water of condensation incidental to said venting to prevent it from entering the food trays or water-receptacle."

MORE PRACTICAL INCANDESCENT LIGHTING

Lewis Howard Latimer

Six years before he was born, Lewis Howard Latimer's parents escaped from slavery in Virginia. At age sixteen, Latimer served in the U.S. Navy during the Civil War. After the war, he acquired a job with a Boston patent law firm and studied drafting. Rising to the position of head draftsman, he helped draft the patent for Alexander Graham Bell's design of the telephone.

This statue of inventor Lewis Howard Latimer was placed in front of Margaret E. Morton Government Center in Bridgeport, Connecticut, in 2015. *Author's collection.*

Later, Latimer was hired as an assistant manager and draftsman by Hiram S. Maxim in Bridgeport, Connecticut. His main task was to help develop an electric light bulb that could compete with Thomas Edison's invention. While working for Maxim, Latimer invented a way to make carbon filament more durable—thus making incandescent lighting more practical. He also invented a threaded socket, an air conditioning unit, a new support for arc lights and a new way to attach the carbonized filament to platinum wires.

In 1884, Latimer became a key employee of Thomas Edison, working on the development of the light bulb. Latimer became widely known as a key witness in Edison's patent infringement lawsuits. He also supervised the installation of public electric lights in New York City, Philadelphia, Montreal and London. He was also the only African American among Edison's closest associates. Latimer was a true Renaissance man—an inventor, draftsman, engineer, musician, poet and author of a book on electricity. He passed away in 1928 at age eighty.

Edison Pioneers was an organization formed on the seventy-first birthday of Thomas Edison by his earliest employees. Its historian, and Edison's private secretary, William H. Meadowcroft remembered fellow member Latimer's "broadmindedness, versatility in the accomplishment of things intellectual and cultural, a linguist, a devoted husband and father, all were characteristic of him."

INSTANT CAMERA

Edwin Herbert Land

Edwin H. Land was born in Bridgeport, Connecticut, in 1909. During his life, he accumulated a total of 533 patents. However, his most important invention by far was that of the instant camera in 1953. His three-year-old daughter asked him why a camera could not immediately print out pictures. That got him to thinking. Soon, Land developed the instant camera. He announced the first instant color photograph system in 1959, and four years later, the camera was available for purchase.

A Polaroid Land Camera. Connecticut-born Edwin H. Land invented these instant cameras.

Born to a working-class family (his father operated a scrap metal and salvage business in Bridgeport), Land attended the prestigious Norwich Academy and became interested in science and technology. Besides the Polaroid instant cameras, Land and his scientists also invented revolutionary "sunglasses, camera filters and equipment for three-dimensional movies." Land retired from his company in 1982 after having served as its chairman of the board and, from 1937 to 1975, its president. He never received a college degree, but he was awarded honorary doctorates by many colleges and universities, including Harvard University.

LITHIUM-ION BATTERY

John B. Goodenough

Nobel Prize winner John Goodenough developed a lithium battery in 1980. *Author's collection.*

In his Nobel Prize autobiographical article, John B. Goodenough remembers growing up seven miles north of downtown New Haven on the main bus route to Waterbury. He mentions his house had "an adjoining woodshed, a large barn, an ice house, and a windmill for pumping our water from a spring in the back lot…[which was] replaced by an electric pump and the coal-fired furnace by an oil furnace."

In childhood, Goodenough suffered from undiagnosed dyslexia. Years later, he wrote, "I never was a good reader, and through my school years, I worked hard to cover my deficiency. I also had a deep sense of insecurity that only lifted slowly as I grew older." However, through aptitude and determination, he managed to complete a Yale University bachelor's degree in mathematics during his service as a U.S. Army Air Forces meteorologist in World War II.

When the war was over, Goodenough was accepted into the graduate program at the University of Chicago. There, he earned a master's degree in physics in 1951 and a doctorate the following year. Upon entering the school, one instructor told him, "I don't understand you veterans. Don't you know that anyone who has ever done anything significant in physics had already done it by the time he was your age; and you want to begin?"

Begin he did, and it led to the creation of the rechargeable, lightweight lithium-ion battery, which makes today's portable computers, smart phones and electric vehicles possible.

After a career that included work at the Massachusetts Institute of Technology (MIT) and the University of Oxford, in 2019, Goodenough shared the Nobel Prize for Chemistry with American chemist M. Stanley Whittingham and Japanese chemist Yoshino Akira. At 97 years old, he was the oldest person to ever win a Nobel Prize in its 118-year-long history.

In its biographical material on Goodenough, the Nobel organization stated, "In 1980 John Goodenough developed a lithium battery with a cathode of cobalt oxide, which, at a molecular level, has spaces that can house lithium ions. This cathode gave a higher voltage than earlier batteries. Goodenough's contributions were crucial for the development of lithium-ion batteries, which are used in for example mobile phones and electric cars."

Today at age one hundred, Goodenough is a professor in the Department of Electrical and Computer Engineering at the University of Texas at Austin and Virginia H. Cockrell Centennial Chair in Engineering.

LOCKS

Linus Yale Jr.

In 1868, inventor Linus Yale Jr. and businessman Henry Robinson Towne founded the Yale Lock Manufacturing Company in Stamford, Connecticut. Linus's father, Middletown, Connecticut–born Linus Sr., was a noted bank lock inventor and manufacturer. The younger Yale followed in his footsteps, making a career of inventing and developing locks. During his life, Linus Jr. held demonstrations for potential customers that highlighted his ability to pick the locks of the time.

Inspired by the ancient Egyptian wooden locks with pin tumblers, Yale's father was awarded a U.S. patent on a new type of pin-tumble lock. When

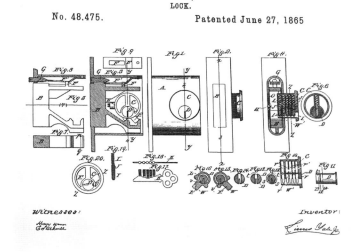

A drawing accompanying Linus Yale Jr.'s padlock patent of 1865. *United States Patent and Trademark Office.*

his father died, Linus Jr. took over management of the family lock shop and proceeded to invent what would become the most popular locks in the world.

At first concentrating on bank safes and vaults, Yale invented a lock that had a combination that could be set by the owner. His locks used flat keys with serrated edges that moved internal pins into a position that allowed the locks to open. During this time, Yale was awarded a series of patents on new designs and improvements in existing designs. His locks, with their different keys matching different combinations, were the first mass-produced product with different objects.

In the twenty-first century, Yale's locks are still the most common type found in homes and businesses. For decades after its founding, Yale's company was based in Stamford, Connecticut. In the mid-twentieth century, it moved operations to Tennessee. In 2000, after its acquisition by the ASSA ABLOY corporation, its headquarters returned to its home state—to Berlin, Connecticut. In 2006, Linus Yale Jr. was inducted into the National Inventors Hall of Fame.

CLOTHES HOOK

O.A. North

In 1869, fifty-year-old Orrin A. North of New Britain was awarded a U.S. patent for a "clothes hook," from which has descended the modern wire coat hanger. North described his innovation as "new and useful Improvements in Coat and Dress-Supporter" and "to enable others skilled in the art to make and use the same."

North described his clothes hook as "formed of a bent metal rod or wire" and stated that if a man has a "choice garment," he will be able to better keep its shape "by placing the arm holes upon the outer ends of the supporter." His clothes hanger could accommodate several garments at a time and could be easily dissembled and assembled.

North also patented improved doorbells (1869), wooden fruit baskets (1872) and basket handles (1872). He died in New Britain in 1884.

HAWAIIAN ALOHA SHIRT

Ellery J. Chun

Hawaii native Ellery J. Chun is credited with popularizing the original Hawaiian aloha shirt. The son of Chinese merchants in Honolulu, he graduated from Yale University in New Haven with a bachelor's degree in economics in 1931. Back home, his family's dry goods business experienced the effects of the Great Depression, and they decided that a new product was needed. Ellery Chun then initiated the first mass production of the colorful Hawaiian shirts and named them "aloha shirts."

Chun and his sister Ethel Chun Lum customized the shirts with pictures of palm trees, ukuleles, surfers, pineapples and other "things Hawaiian." They were incredibly popular with tourists. Using cloth imported from the U.S. mainland, Japan, China and Tahiti, he had the shirts manufactured in the Hawaiian Islands and distributed throughout the world. In 1936, he registered the "Aloha Shirt" trademark.

Years later, he left retailing, closed his company and moved into banking. However, other manufacturers continued to profit from the popularity of the Hawaiian shirt. Chun passed away in 2000 at age ninety-one.

Today, aloha shirts are worn by most workers in Hawaii on Fridays— including on the floor of the Hawaii State House and Senate.

CABBAGE PATCH KIDS

The Greenberg Family

Coleco Industries, Inc. was an American corporation founded in 1932 by Russian immigrant Maurice Greenberg. Under its original name, the Connecticut Leather Company, it sold shoe repair supplies to shoemakers. Maurice's sons Leonard and Arnold Greenberg eventually became in succession Coleco's presidents and CEOs and then chairmen of the board of directors.

Founder Maurice Greenberg, a resident of West Hartford, died at age seventy-nine in 1980. His sons diversified the company into other product lines, including toys, electronic games, sports equipment and aboveground swimming pools. Through these years, its corporate headquarters were located on Quaker Lane South in West Hartford, Connecticut.

Products during the 1980s included ColecoVision, a home video game console released in 1982; however, Coleco's most successful products of all time were the Cabbage Patch Kids dolls.

One of its biggest disappointments was the 3.58 MHz Adam computer, which was released in 1983 and was only sold for a couple of years. It came with 80 KB RAM, 16 KB video RAM and 32 KB ROM. It was priced at about $700 and included a printer but no monitor. (Like most of its competitors, it required a television set for visual display.) A choice of 20-, 30- or 40-megabyte hard disk drives were available—not bad at the time, but it pales in comparison to twenty-first-century personal computers with their $60 external drives with capacities measured in terabytes.

Partly as a result of problems with Adam computer sales, in the late 1980s, Coleco sold off its assets and went out of business. Today, some 1980s Cabbage Patch Kids dolls can be worth over $1,000 to collectors.

DIAPERS

Marion Donovan

On January 19, 1949, Marion O'Brien Donovan of the Saugatuck part of Westport, Connecticut, applied for a patent on her leak-proof diapers. As she stated in her application, "diapers and covers were not leak-proof and therefore accessories such as mattress covers, rubber sheets, rubber pants and similar articles were needed." In 1975, she told interviewer Barbara Walters that she had asked herself, "What do I think will help a lot of people and most certainly will help me?"

Patent number 2,556,800 was granted on June 12, 1951. Within months, Donovan sold the rights to this patent to Keko Corporation for $1 million. Donovan's diapers "used metal and plastic snap fasteners instead of safety pins, were covered with waterproof and reusable nylon parachute cloth, and the insides were made of cloth to allow the baby's skin to breathe." She named it the Boater because she thought it looked like a boat.

After receiving the patent, Donovan began work on the creation of a disposable paper diaper. That was ten years before Pampers were introduced. However, no manufacturers at the time were interested in Donovan's invention.

After college, Donovan worked as an editor for *Vogue* magazine before leaving to raise her three children. She later patented several products, marketed and/or manufactured them herself and in 1958 earned a master's degree in architecture from Yale University.

Chapter 2

FOOD

B eginning with colonial cooking, which was influenced by Native American cooking, Connecticut has been one of the culinary centers of the United States. At the end of the eighteenth century, resident Amelia Simmons wrote the first cookbook that was printed and sold in America. In the nineteenth and twentieth centuries, the influx of people from the American South, Europe, Central and South America, Asia and Africa brought the widest possible selection of food choices.

FIRST AMERICAN COOKBOOK

Amelia Simmons

In 1796, Amelia Simmons self-published, at her own expense, *American Cookery*, which is the first cookbook printed and sold in America. That was at Hartford's Isaac Beer's Bookstore. Simmons's book gives a glimpse into colonial American culture and was dedicated to the "improvement of the rising generation of Females in America." However, its first edition was a disappointment—the first thirteen pages contained places to buy products, and her "entrusted" recipes had errors in both cook times and ingredient measurements. Her second printing, published the same year, was error-free and included errata; for example, "Page 25. Rice pudding No 2; for one pound butter, read half pound—for 14 eggs read 8." She explained all endorsements were "without her consent—which was unknown to her, till after publication." Simmons hired a person who omitted "several articles very essential…and placed others in their stead" because of "not having an education sufficient to prepare the work for press." Simmons's book was published four times from 1796 until the 1830s; it gave practical, helpful advice to the reader.

Titled *American Cookery, or the Art of Dressing Viands [Meat], Fish, Poultry and Vegetables, and the Best Modes of Making Pastes, Puffs, Pies, Tarts, Puddings, Custards and Preserves, and All Kinds of Cakes from the Imperial Plumb to Plain Cake: Adapted to This Country, and All Grades of Life*, it combined British cooking and the more readily available American ingredients. Simmons states, "This being an original work in this country," we see words like "receipts" (recipes), "slaw," "cookie" and "pompkin." Some recipes in the book use molasses instead of the more expensive West Indian sugar. She includes a larger number of eggs since eighteenth-century chickens and eggs were much smaller than those of today.

Simmons was an orphan and wanted to give "general and universal knowledge [to] females in this country, who by the loss of their parents" would have the benefit of her experience and practical knowledge on how to bake, pickle, preserve and be able to select the freshest meat, fish, vegetables and dairy products so as to avoid swindlers who use "deceits…to give them a freshness of appearance." She explains eighteenth-century herbs in food preparation; she instructs how to "dress" meat, make stuffing and prepare some recipes like stews, pies, puddings, custards, biscuits and rusks (dry, hard bread, used today for baby's teething food).

Her native recipes include using the plentiful Indian meal or cornmeal, like her Johnny cakes and Indian slapjacks. Simmons's puddings are made with fruits or vegetables like winter squash or pumpkin, which is similar to our pumpkin pie. Her recipes contained the flavoring of rose water or heavier spices, like in her gingerbread recipes, or her use of hard and sweet "cyder," beer or wine.

Simmons's recipes are still used by colonial enthusiasts who try to replicate them in their original form. Following are a few:

Johny Cake or Hoe Cake. Scald 1 pint of milk and put to 3 pints of Indian meal, and half pint of flower [sic]—bake before the fire. Or scald with milk two thirds of the Indian meal, or wet two thirds with boiling water, add salt, molasses and shortening and work up with cold water pretty stiff, and bake as above.

Pompkin No. 1. One quart stewed and strained [pumpkin], *3 pints cream, 9 beaten eggs, sugar, mace, nutmeg and ginger, laid into paste* [pastry] *No. 7 or 3, and with a dough spur, cross and chequer it, and baked in dishes three quarters of an hour.*
No. 2. One quart milk, 1 pint pompkin, 4 eggs, molasses, allspice and ginger in a crust, bake 1 hour.

Cookies. One pound sugar boiled slowly in half pint water, scum well and cool, add two tea spoons pearl ash dissolved in milk, then two and half pounds flour, rub in 4 ounces butter and two large spoons finely powdered coriander seed, wet with above; make rolls half an inch thick and cut to the shape you please; bake fifteen or twenty minutes in a slack oven—good three weeks.

To make a fine Syllabub from the Cow. Sweeten a quart of cyder with double refined sugar, grate nutmeg into it, then milk your cow into your liquor [liquid], *when you have thus added what quantity of milk you think proper, pour half a pint or more, in proportion to the quantity of syllabub you make, of the sweetest cream you can get all over it.*

A Sick Bed Custard. Scald a quart milk, sweeten and salt a little, whip 3 eggs and stir in, bake on coals in a pewter vessel.

Carrot Pudding. A coffee cup full of boiled and strained carrots, 5 eggs, 2 ounces sugar and butter each, cinnamon and rosewater to your taste. Bake in a deep dish without paste.

Amelia Simmons and the First Cupcake

The earliest mention of a cupcake was in Amelia Simmons's 1796 self-published cookbook, *American Cookery*. She called it "a light cake to bake in small cups." The eighteenth-century cook appreciated this time-saving technique since the "small cup cakes" cooked much faster in individual ramekins. A full-size cake took hours to make since the cook needed to tend to the fire while judging and determining when the cake was fully done.

In Simmons's cupcake recipe, she uses "emptins," which is short for "empty-ings," a liquid yeast that remains at the bottom of beer barrels, cider or wine containers. She includes a recipe on how housewives or cooks can make "emptins."

The cook spent much time in making a cake rise by beating egg whites or eggs by hand strenuously for two to three hours and using twelve to thirty-five eggs (egg size was smaller then) or using a bread yeast in a cake and then waiting several hours for the cake to rise—or making a smaller cake (what we call a cookie), which used fewer eggs for rising. Simmons's "cup cake" recipe below provided another alternative; it uses "emptins" as a leavening agent. She uses the flavoring of rosewater in both recipes. Rosewater is said to have a calming, mood-lifting aroma besides aiding the digestive system.

> *A light Cake to bake in small cups: Half a pound sugar, half a pound butter, rubbed into two pounds flour, add one glass wine, one do. rosewater, two do. emptins, a nutmeg, cinnamon and currants.*
>
> *Soft Cakes in little pans: One and half pound sugar, half pound butter, rubbed into two pounds flour, add one glass wine, one do. rose water, 18 eggs and a nutmeg.*
>
> *Emptins: Take a handful of hops and about three quarts of water, let it boil about fifteen minutes, then make a thickening as you do for starch, strain the liquor, when cold put a little emptins to work them, they will keep well cork'd in a bottle five or six weeks.*
>
> *Gingerbread Cakes, or butter and sugar Gingerbread: Three pounds of flour, a grated nutmeg, two ounces ginger, one pound sugar, three small spoons pearl ash dissolved in cream, one pound butter, four eggs, knead it stiff, shape it to your fancy, bake 15 minutes.*

HOME ECONOMICS FOUNDER AND AMERICA'S FIRST CELEBRITY COOK

Maria Parloa

Cookbook author and cooking instructor Maria Parloa was a founder of the science of home economics and is considered by many to be America's first celebrity cook. She was also likely the most famous resident of the Connecticut town of Bethel. Orphaned at an early age, she supported herself by cooking at New Hampshire summer resorts. In 1872, at age twenty-nine, she published *The Appledore Cook Book*, which was named after one of the resorts. One of its recipes is said to be first recipe for tomato soup ever published.

In the introduction to her cookbook, Parloa notes, "The great trouble with all the cook books which I have known (and I am constantly hearing the same complaint) is, that they are too expensive, and that they use weight instead of measure, and also that they take for granted that the young housekeeper knows many things which she really does not."

One of the most popular cookbooks of the nineteenth century, *The Appledore Cook Book*'s sections were organized in two parts—the first for "plain cooking" and the second for "rich cooking." The plain cooking included fish, soups, meats, vegetables, bread, plain cake, puddings and pies. The part on rich cooking had sections on soups, fish, poultry, venison, entremets, puddings, pies, pudding sauces, dishes for the sick, desserts, cake, preserves, pickles, sauces, drinks, eggs and miscellaneous.

Four years after her cookbook's release, Parloa gave her first public speech—in New London, Connecticut—which raised money for a Sunday school organ. That led to many other talks, including in Boston, where she also opened a cooking school. Ultimately, fees from her public lectures led to her becoming very wealthy. In 1878, she visited England and France, where she attended cooking classes. From these experiences, the following year she published *First Principles of Household Management and Cookery*. As one of the first instructors of the Boston Cooking School, she enabled women to enter upon a career in cooking. She also taught courses for Harvard medical students on sick-room cookery. In 1883, she opened a cooking school in New York City and allowed many immigrant women to attend classes for free.

Maria Parloa's books include *Miss Parloa's New Cook Book: A Guide to Marketing and Cooking* (1881), *Practical Cookery with Demonstrations* (1884), *Miss Parloa's Young Housekeeper: Designed Especially to Aid Beginners* (1893), *Home Economics: A Guide to Household Management, Including the Proper Treatment of the Materials Entering into the Construction and Furnishing of the House* (1898) and *Chocolate and Cocoa Recipes by Miss Parloa, and Home Made Candy Recipes by Mrs. Janet McKenzie Hill* (1909).

Parloa was a key member of the group that founded the American Home Economics Association, which became the most important professional association for home economists. She authored articles for the *Ladies' Home Journal* and two Farmers' Bulletins for the U.S. Department of Agriculture. She spent the last years of her life at her home in Bethel, Connecticut. At her passing, she left her private library and enough money to found the Bethel Public Library, which today honors her memory with its Maria Parloa Community Room.

GRIDIRON

Amasa and George Sizer

Most commonly, a gridiron is either a football field or a grate for broiling food. The latter was what Amasa and George Sizer of Meriden, Connecticut, invented in 1836. Together, the men invented a useful cooking device that combined a gridiron and a spider into one instrument, but it could be used together or separately. The spider, which they illustrated in their patent, had four legs and a handle extending from the pan and was used to catch the juices of the broiled meat. When it was "folded it could be used as a chafing-dish." The gridiron, a metal grate that had parallel bars (to grill food like fish, meat and vegetables), was placed on top of the spider and was attached so the gridiron could be raised or lowered. Both units were made out of any metal, but the inventors' preference was using cast iron.

Half a century later, in 1889, William C. Perkins of the New Haven Wire Goods Company received the patent for a hinged gridiron. He combined two steel wire grids that were hinged to fold together, and the food would be held securely while grilling over an open flame. The two hinged wire grids clamped together prevented food from falling into the fire. Moreover, each grid could face the fire so both sides could be grilled evenly.

Exactly fifty years later, in 1939, Caesar Rossini and Tazio Pieragostini of New Haven were awarded a patent for a hinged gridiron that was automatically adjustable. Their manufactured hinged gridiron or broiler was for commercial use. The handle to open and close the double-hinged gridiron also had a mechanism that clamped both wire grids together. It was hinged on one side, and grilling took place in a vertical position; however, it swung to a horizontal position to release its broiled food. The rear end of the two wire grids was permanently clamped but would move to adjust itself to the thickness of the food being grilled. Louis' Lunch, also located in New Haven, still broils its hamburgers using a vertical gridiron.

LOLLIPOPS

George P. Smith

In 1908, New Havener George Smith created a hard candy on a stick and named it Lolly Pop. Shortly afterward, one of his plant supervisors, Max

Buchmuller, invented a machine to insert the candy onto the sticks. His 1909 patent application, titled "Candy-Making Machine," stated, "This invention is in the candy making art, and has for its object to provide improvements in the manufacture of lolly pops, to the forming and molding of the candy, and to the inserting of the sticks into the formed or molded candy while this is held in position in a conveyer."

This patent was approved—but not until over five years later, in 1914. Seventeen years after that, the Bradley Smith Company received a trademark for the name "lollipop." Today, the number of Lolly Pops (aka lollipops) sold annually in the United States is well into the billions.

In April 2022, lawmakers in the Connecticut state House of Representatives approved a bill that included the designation of the lollipop as Connecticut's state candy.

Hamburgers

Louis' Lunch and Jack's Lunch

Louis' Lunch, a restaurant located in New Haven, Connecticut, is recognized by the Library of Congress as the birthplace of the first hamburger sandwich, although some states dispute this.

Danish immigrant Louis Lasson was a hard worker who bought a wooden cart in 1895 and sold lunches to New Haven people who worked in nearby factories. This became Louis' Lunch Wagon from 1907 to 1916. In the twentieth century, many popular, movable food stands transitioned into stationary restaurants. Louis' Lunch was one of them.

In 1900, Lasson wanted to serve a customer who was in a rush to leave. Using his own blend of steak trimmings, he quickly made a beef patty, grilled it, put the meat between two pieces of toasted bread and left a satisfied customer with a meal that he could eat on the go. This carryout food began a new American tradition—Louis is credited with creating the first hamburger sandwich. Soon busy locals, tourists and students and workers from Yale University patronized Louis' Lunch, dining on tasty, freshly grilled-to-order hamburger sandwiches.

The success of Lasson's product, the hamburger, might also be due to its rise in popularity after the "burger-on-a-bun" was introduced to millions of people at the 1904 St. Louis World's Fair, which was also known as the Louisiana Purchase Exposition.

Louis' Lunch became a favorite and popular place to eat. That was very fortunate since in the 1970s, the building appeared to be doomed by a high-rise construction. A safe haven on Crown Street was finally found for the historic building—supposedly hours before its demolition deadline. After making a thirty-minute ride aboard a truck from its original location, the historic little building badly needed reconstruction. Today, Louis' Lunch still sits on Crown Street with many new bricks from around the globe incorporated into its historic walls as a silent testimony of the generosity of its friends and supporters.

There is a much older history surrounding the hamburger sandwich. The word *sandwich* originated in 1762, when John Montagu, the fourth Earl of Sandwich, wanted to quickly eat something. It is thought that he didn't want to leave the gambling tables, so he is credited with the word *sandwich*, which became totally synonymous with totally portable food. Hamburger has no ham in it but instead comes from the term "Hamburg steak." Hamburg, Germany, was known for its superior beef, which came from high-quality cows that were raised in its countryside. In the late nineteenth century, hamburger steaks became popular in the United States along with minced beef and onion patties that were served to customers without the bread. It seems logical that we can put the two together and create the hamburger sandwich, which Louis' Lunch did in the past and still does today.

The first steamed cheeseburger is credited to have been invented by Jack Fitzgerald in his Middletown, Connecticut restaurant during the 1920s. Jack's Lunch, located on Main Street, was extremely popular with locals who worked in the center of town and students from nearby Wesleyan University. Jack sold steamed cheeseburgers in eighteen-tin trays filled with ground beef patties that were cooked in a "tall, copper box filled with simmering water." Unlike Louis' Lunch, Jack's Lunch closed its doors after forty-four years. Today, the hamburger still reigns supreme as one of the world's most popular foods, with approximately fifty billion eaten in the United States alone each year.

The uniqueness of Louis' Lunch has been recognized on television on the Travel Channel and the Food Network, in print in *Food & Wine* magazine and by word of mouth. Its proximity to Yale University accounts for some of this attention. Louis' Lunch may look small amid some of its taller neighbors, but it still follows its traditions—grinding five cuts of fresh meat daily and freshly cooking hamburgers to order on their 1898 original gas-powered, cast-iron grills that cook the meat patties vertically.

The fourth-generation family-owned Louis' Lunch keeps it simple, like the original proprietor, placing the hamburger only on white toast with the "meat speaking for itself." They serve three trimmings on their hand-rolled hamburger: melted sharp cheddar cheese, griddled onions and ripe tomato slices. No other condiments like ketchup or mustard are provided. Once entering their brick red front door, not much has changed since their prototype hamburger began. Unless instructed otherwise, the beef is cooked medium rare. Both regulars and guests still enjoy Louis' quaint, old-fashioned dining experience or order its food as takeout. The menu is simple: an original burger "grilled to perfection," potato salad, chips, homemade pie and a nonalcoholic drink.

Peter Paul Candy

Peter Halajian

In the 1980s, Connecticut's Peter Paul, Inc. manufactured two of the most popular candies in the country: Mounds, with dark chocolate covering shredded coconut, and Almond Joy, which was basically a milk chocolate Mounds bar topped with nuts. As if the two bars weren't already known by everyone in America, in 1977, the company released one of the most unforgettable advertising jingles in history: "Sometimes you feel like a nut, sometimes you don't."

Peter Paul founder Peter Halajian immigrated to the United States from Armenia in 1890 when in his mid-twenties. Later, he changed his last name to Paul. After earning money as a Naugatuck factory worker for a few years and

Popular candy bars from the Peter Paul and Mars companies. *Author's collection.*

selling candy on the side, he opened his own store—a candy shop. Within a few years, two more shops followed in Torrington and Naugatuck. By 1919, he had convinced six other Armenian immigrants to start a larger business—the Peter Paul Manufacturing Company. They introduced the Mounds candy bar the following year and in 1946 invented the Almond Joy bar.

The company's products remained popular throughout the Great Depression, when the company took the counterintuitive step of doubling the size of its product while keeping its price the same. Sales increased so much that profits outweighed the additional raw material costs, and the company was able to build an addition to its Naugatuck plant and add to its workforce. All at the height of the Depression!

During World War II, most of Peter Paul candy bars production was sent to U.S. troops, leaving little remaining for the consumers back home. Also, it lost access to its coconut suppliers when Japanese forces occupied the Philippines. This necessitated a quick—and successful—switch to Caribbean sources. Over all this time, Peter Paul retained its Naugatuck factory—even after the company was sold to British candy maker Cadbury Schweppes in 1978. However, ten years later, Hershey's bought Peter Paul and in 2007 moved its Connecticut operations to Virginia.

MARS CANDY

Forrest Mars Sr.

While, unlike Peter Paul, Mars, Inc. was never headquartered in Connecticut, it was a Connecticut-educated man who was responsible for its product line.

An industrial engineering graduate of Yale University in 1928, Forrest Edward Mars Sr. was the son of Mars, Inc. founder Frank C. Mars. As a teenager, Forrest helped come up with the idea of "malted milk in a candy bar." In 1923, it would be released under the name Milky Way. Later, Forrest was responsible for introducing some of the most popular chocolate candy bars in the world: Snickers in 1930 and 3 Musketeers in 1932. In the early 1940s, he oversaw what is arguably the most popular Mars candy: M&Ms. Like Peter Paul with its catchy jingle, Mars had one for its hard-shelled M&Ms: "melt in your mouth, not in your hand."

BIRTHPLACE OF THE OREO COOKIE AND BARNUM'S ANIMAL CRACKERS

Adolphus W. Green

Many people believe that the Oreo cookie was founded at a Victorian mansion in Greenwich, Connecticut. *Author's collection.*

A black-and-white Victorian mansion in Greenwich has been called the birthplace of the Oreo cookie and Barnum's Animal Crackers. It was owned by Adolphus W. Green, who was chairman and president of the National Biscuit Company (since renamed Nabisco), as a vacation home and built in 1886. Green rented the home for a few summers and purchased it in 1905. It was while living there that Green oversaw the creation of the Oreo cookie and Barnum's Animal Crackers. In 1912, the Oreo cookie was first made in bakeries at Chelsea Market in New York City. Green owned the house until his death in 1917. His funeral was held at St. Mary's Catholic Church in Greenwich.

Today's ten-thousand-square-foot mansion has a totally renovated interior that would be unrecognizable to Green, but the outside is basically unchanged since Green lived in it. With a turret the color of an Oreo, it is thought that its resemblance to the chocolate-and-vanilla creme cookie was intentional.

Since its first sale, almost half a trillion Oreo cookies have been sold.

LOBSTER ROLLS

Perry's Restaurant

The hot lobster roll was invented in Connecticut. Although there is dispute regarding the exact time and location, the first official documentation shows it was in 1929 at Perry's Meat/Fish Market in Milford, Connecticut. A "traveling liquor salesman" wanted a "hot, grilled lobster sandwich" to take with him, so quick-thinking Harry Perry made one for him to avoid losing a sale. (Perhaps this salesman's request came from a past memory of when thrifty New England lobstermen took their unsold day's catch and cooked it up for a quick sale.) Perry thought he could increase his profits if he improved on his hot lobster sandwich. His impromptu sandwich didn't hold the moist lobster meat well, and the sliced bread was flimsy, tasteless and soggy. He wanted his sandwich to hold his lobster and not dissolve. Perry teamed up with French's Bakery in Bridgeport, Connecticut, which created a baked roll similar in shape to a submarine sandwich. After a V-shaped hole was cut into the top and the wedge of bread was removed, Perry added his hot lobster meat, replaced the bread and grilled the entire sandwich in butter. Connecticut's hot lobster roll was born.

Because today's hot lobster roll is so popular, it's almost inconceivable that lobsters were once called "cockroaches of the sea," but back in the 1700s, they were so plentiful that these crustaceans would wash ashore in one- or two-foot-high piles. Frugal New Englanders used them as fertilizer, and because this food was so abundant, it was considered to be a poor man's meal or sometimes a "poor man's chicken." Employees, who had their meals included in their workplace, stipulated that their employer could only serve lobsters two or three times a week. One reason for their dissatisfaction was likely caused by the food's quick deterioration. Lobster meat tastes best when it's fresh. However, back in the eighteenth century, a lobster was cooked after it died, and most did not realize the taste of lobster meat would greatly improve if it was freshly cooked.

During the nineteenth century, lobster became more popular. Travelers either didn't know or didn't care about lobster's negative stigma. Northeast train passengers enjoyed lobster when it was on the menu, and it was equally beneficial with train owners, as lobsters were easy to prepare and inexpensive to get. During the food shortage in World War II, entrepreneurs provided canned lobster meat to U.S. soldiers and to the consumer—both rich and poor—because it wasn't rationed. In the 1950s, the hot lobster roll was so popular that lobster shacks popped up along automotive routes; some turned into year-round restaurants.

Connecticut's inventor did not cover up this briny crustacean's meat; it was left naked. That's why the Connecticut style is sometimes called a "naked" lobster roll. There is no filler, no fancy ingredients—just a grilled roll drizzled with warm butter and topped with fresh lobster meat. Connecticut's hot lobster roll not only tastes good but is also good for you! Lobster meat is filled with lean protein, amino acids, phosphorus and calcium, promoting healthy functions of the body. Many people like the easy convenience of eating Connecticut's hot lobster rolls that are quickly made.

Today, nearly one hundred years after its invention, there is much competition in the Nutmeg State on who serves the tastiest and best hot lobster roll. It can be local restaurants, summer eateries, food trucks or beachside lobster shacks. All agree that the Connecticut hot lobster roll is a delicious treat—prized for its warm lobster morsels that are placed inside a straight-up-sides, buttered, toasted New England split-top bun. Some eateries will seriously weigh out seven or eight ounces of succulent lobster meat, especially if it includes the more expensive tail meat that's broken into large chunks. Other hearty lobster rolls can contain almost two lobsters. Sometimes the price isn't even listed on the menu; instead, the cost is noted at "today's market price." However, the majority of Connecticut's hot lobster rolls are made with a combination of claw or leg meat and sweet, tender knuckle meat.

The grilled buttery hot lobster sandwich has widespread appeal, and many look forward to summertime, when it's in season. This delicious specialty is even more enjoyable if eaten outside while looking at a summer scenic view from any vantage point along Connecticut's long shoreline, river or while looking at a beautiful sunset. Whether you are a native Connecticuter or a vacationer, you can count on the fact that Connecticut's invention of the hot lobster roll is here to stay.

Bigelow Tea

Ruth Bigelow

In 1945, inspired by recipes from colonial America, forty-nine-year-old interior decorator Ruth Campbell Bigelow decided to create an improved cup of tea. Testing out various tea blends and flavors in the kitchen of her New York City home, she developed the first specialty tea in the United States. A unique mixture of several black Ceylon teas, flavored with orange peel and spices, it received many positive comments from friends. So, Ruth named it Constant Comment. Today, it is one of the most popular teas in America.

Ruth and her businessman husband, David Bigelow Sr., began their own company, selling Constant Comment as well as other teas that she developed. Ruth personally approached the owners of small stores and requested they carry her teas. One of them told her he had been opening up a container of her tea for customers "to enjoy the tea's amazing aroma." He continued,

Bigelow Tea founder Ruth Bigelow with her husband, David Bigelow Sr., and their son, David Bigelow Jr. *Bigelow Tea.*

"One whiff and they were sold!" Ruth proceeded to add to each case of tea that she sold a small free jar of Constant Comment with a label that read "open and whiff." Since the family could afford only a one-colored label, many nights both David Jr. and Sr. hand-painted the tea labels before loading up their station wagon to make store deliveries. In the early 1950s, the Bigelow family moved their company to Norwalk, Connecticut.

Their ten-year-old business had a huge setback and financial loss when two hurricanes caused a devastating flood that hit their building and washed their tea downriver. Ruth and David, both over sixty years old, "learned two lessons that we still adhere to: always pay your bills on time, and when you get knocked down, just get right back up!" After several months, they were operational. The company later moved to Fairfield, Connecticut, where its headquarters is still located today.

In the 1960s, David Bigelow Jr. was elected president of the company. He and his wife, Eunice, decided to sell their Bigelow teas in grocery stores nationwide so virtually every shopper had access to Bigelow tea. They also introduced a foil pouch that protects the tea bags' freshness and flavor. It became the number-one specialty tea company in the country.

Today, 100 percent family-owned Bigelow Tea markets 150 varieties of flavored, traditional, green, herbal, decaffeinated, steep by Bigelow Organic, Bigelow Benefits Wellness and Bigelow Botanicals Cold Water Infusions. Both David Jr. and Eunice still serve as co-chairs of the company and personally taste teas before they are produced to maintain high standards of quality control.

Currently, after tea is first blended in Connecticut, it moves to Bigelow's Boise, Idaho manufacturing plant to be processed with its automated, custom-made machines that are responsible for weighing the precise amount of tea that is folded into a paper filter and tied with a thread and tag, sealed in a foil packet and then boxed and shipped throughout the country.

In 2005, David Jr. and Eunice's daughter Cindi Bigelow became president and CEO. She further expanded the reach of Bigelow tea by using social media and new online marketing tools. Cindi has seen her family's company become one of the first in Connecticut to install solar panels, and she introduced innovations that have given Bigelow Tea the distinction of being designated a "zero waste to landfill" company.

Lori Bigelow, Ruth and David's other daughter and a master tea blender, started learning her craft in her parents' kitchen during the 1960s in their Westport, Connecticut home as she assisted her parents in tasting tea after returning home from school. She is credited with making some of Bigelow's favorite tea blends. Among many other accomplishments, Lori was "instrumental in Bigelow Tea's purchase and restoration of the Charleston Tea Garden," which is located on historic Wadmalaw Island, south of Charleston, South Carolina. It is America's largest working tea farm, having an ideal climate for growing tea "the old-fashioned way"—similar tea farms are located thousands of miles away in other parts of the world. The tea farm's visitor center is open and gives factory tours. The public may learn how some of their bushes are descended from 1700s tea plants, how the garden uses cuttings rather than planting seeds, how harvesting

Ruth Bigelow invented Constant Comment tea in the 1940s. Today, her company is based in Fairfield, Connecticut, where its tea is blended. *Bigelow Tea.*

tea takes place about every fifteen days and that quality tea is bought by taste rather than weight.

Today, Bigelow's Constant Comment variety is still one of its best sellers. Perhaps important to that status is the fact that David Jr. and Eunice Bigelow still taste each new batch of Constant Comment. From the creation of one popular original tea, this three-generation family-owned business grew and expanded during its seventy-five-year history. Today, while producing two billion tea bags annually, they are still passionate about creating a perfect cup of tea.

NEW HAVEN–STYLE PIZZA

Frank Pepe

Emigrating from Italy in 1909, Frank Pepe became nationally famous for his pizza, which was ultimately dubbed "New Haven–style pizza." His thin-crusted pizza was similar to what he grew up with near Naples, Italy, and caught on quickly with both fellow Italian immigrants as well as the general American population.

Frank's wife, Filomena Pepe, was a great help to him in making his business venture a success. On the official Frank Pepe website, it states Filomena was "a partner to Frank and his greatest support. Unlike her husband, Filomena could speak and write in English so she was invaluable in establishing and developing the business since she took care of all of the administrative tasks and other things requiring such knowledge.

In 1925, Pepe opened Frank Pepe Pizzeria Napoletana on Wooster Street in New Haven. Later, the restaurant moved into a larger building next door. Still later, both buildings were part of the restaurant. Many celebrities have visited Pepe's over the years, including actors Robert DeNiro and Paul Giamatti and the great New York Yankees catcher and manager Yogi Berra.

Today, in addition to Pepe's original location on Wooster Street in New Haven, there are six other Pepe's in Connecticut, as well as three in Massachusetts and one each in New York, Maryland and Rhode Island.

PEPPERIDGE FARM

Margaret Rudkin

Born in New York City in 1897, Margaret Fogarty married Wall Street broker Henry Rudkin in 1923. Six years later, after the loss of most of their savings in the stock market crash of 1929, they purchased a home in Fairfield, Connecticut, and named it Pepperidge Farm after an old pepperidge tree on the property. With a family of three children, Margaret took up baking. In an effort to bake healthy all-natural bread, she baked a first loaf that, in Margaret's words, "should have been sent to the Smithsonian Institution as a sample of Stone Age bread, for it was hard as a rock and about one-inch-high." After trial and error, she finally came up with a recipe that was praised by everyone who tried it.

Naming the bread Pepperidge Farm, she tried to peddle it to a local grocery store owner, but he was reluctant to take on a product she had naively priced too high. Only after tasting it was he won over. Once New York City specialty stores agreed to carry her bread, Margaret's husband would take twenty-four loaves each day on his train commute into New York City. The sales began climbing and didn't stop, even though Margaret's bread was priced at twenty-five cents a loaf, while the average price of similar bread was only ten cents a loaf. They soon moved production from the Rudkin garage to a factory building, and in only a few years, they marked the 1,000,000th loaf sold.

During World War II, Pepperidge Farm sales dropped when high-quality ingredients weren't available. But shortly after the end of the war, they opened a new bakery in Norwalk, Connecticut.

Through the 1940s and 1950s, Margaret Rudkin oversaw every aspect of Pepperidge Farm's business, and she became famous as one of the most successful businesswomen in America. By 1961, it was the largest independent baker in the United States, annually selling $40 million of baked goods, including some favorites like Milano®, Brussels® and Goldfish® crackers. In 1961, with 1,500 employees at six facilities, it was acquired by the Campbell Soup Company. Margaret Rudkin was allowed to continue as president of Pepperidge Farm and became the first female director on Campbell's board of directors.

In 1963, the Margaret Rudkin Pepperidge Farm Cookbook became a best seller. Margaret Rudkin passed away in New Haven in 1967 at age sixty-nine.

SUBWAY RESTAURANTS

Fred DeLuca

Starting with its first store in Bridgeport, Connecticut, in 1965, Subway rose by 2002 to have more restaurants than another company in the United States. Today, with corporate headquarters in Milford, Subway has more than forty thousand independently owned restaurants in over one hundred countries.

Founder by teenager Fred DeLuca, the first location was named Pete's Super Submarines after friend Peter Buck loaned him startup money. The name was changed to "Subway" in 1968. In 1974, they owned sixteen Subways in Connecticut and knew they would not meet a goal they had set thirty-two stores. Their solution was to franchise restaurants. They did, and it resulted in an era of incredible growth.

Chapter 3

HEALTH AND MEDICINE

Since its earliest days, Connecticut has been at the forefront in the latest advances in medical care. Yale New Haven Hospital, founded in 1826, and Hartford Hospital, founded in 1854, are two of the major teaching hospitals in the United States. Yale University's medical school professors have participated in countless inventions and innovations.

AMERICA'S FIRST DOCTOR OF MEDICINE DEGREE

Daniel Turner

In 1723, Yale College in New Haven awarded its first doctor of medicine degree. It was also the first given by a college in Britain's North American colonies. The recipient was London, England physician Daniel Turner (1667–1741). He was the author of 1714's *De Morbis Cutaneis*, which was the first English-language dermatology textbook. (At the time, most medical textbooks were in Latin.) He was also someone who made a donation of books to Yale College as he requested the degree.

However, Turner was no ordinary physician—he was a prolific author with over thirty texts and treatises to his name. In addition to his dermatology textbook, his two-volume *The Art of Surgery*, published in

1721, provides perhaps the best information in existence on the state of surgery in the early to mid-1700s. The first edition's title page states, "In which is laid down Such a general idea of the same [that is, surgery], as is founded upon Reason, confirm'd by Practice, and farther illustrated with many singular and rare cases."

The Art of Surgery includes 110 case studies of people from all walks of life—the wealthy, tradesmen, servants, military personnel and poor people. Their ailments include accidents from sports, people kicked or thrown by horses, injuries from sword fights, inflammations and one woman who dislocated her jaw by yawning.

Even though Dr. Turner may have given Yale a little incentive to be granted a doctorate, perhaps the university—and the medical profession—owes him a debt of gratitude for the priceless detailed information he provided to posterity on the state of medicine in his time.

FIRST SCHOOL OF PUBLIC HEALTH

William H. Welch

Born in Norfolk, Connecticut, in 1850, medical doctor William H. Welch established the first school of public health in the United States in 1916— the Johns Hopkins School of Hygiene and Public Health. For one-half a century, Welch was one of the most prominent leaders in American medicine as a physician, a pathologist and a medical school administrator. He was the first dean of the Johns Hopkins School of Medicine and introduced a medical school curriculum that emphasized study of physical sciences and laboratory work.

Welch was also credited with the discovery of *Bacillus welchii*, which is the organism that causes gas gangrene.

In 1870, Welch received an AB degree from Yale University and, in 1875, an MD from the Columbia University College of Physicians and Surgeons.

The William H. Welch Medical Library at John Hopkins is named for Welch.

Pain-Free Dentistry

Horace Wells

In the fall of 1844, twenty-nine-year-old dentist Horace Wells and his wife attended a demonstration of laughing gas (aka nitrous oxide) in Hartford, Connecticut, by showman Gardner Colton. Afterward, Wells asked Colton to administer it to him as he had an infected wisdom tooth extracted. The result was almost no pain from a procedure that would ordinarily have been one of the most painful. Wells proceeded to test nitrous oxide gas on his patients with the same results.

Three years later, Wells published a booklet, "History of the Discovery of the Application of Nitrous Oxide Gas, Ether, and other Vapors to Surgical Operations." In it, he presented evidence that he had discovered the pain prevention properties of various gasses. In the book, Wells noted instances when pain might not be felt:

> *Surgical operations might be performed without pain, by the fact, that an individual, when much excited from ordinary causes, may receive severe wounds without manifesting the least pain; as, for instance, the man who is engaged in combat may have a limb severed from his body, after which he testifies, that it was attended with no pain at the time: and so the man who is intoxicated with spirituous liquor may be severely beaten without his manifesting pain, and his frame, in this state, seems to be more tenacious of life than under ordinary circumstances.*

Wells continued:

> *By these facts I was led to enquire if the same result would not follow by the inhalation of exhilarating gas, the effects of which would pass off immediately, leaving the system none the worse for its use. I accordingly procured some nitrous oxide gas, resolving to make the first experiment on myself, by having a tooth extracted, which was done without any painful sensations. I then performed the same operation for twelve or fifteen others, with the like results.*

In 1864, the American Dental Association posthumously recognized Wells as the discoverer of modern anesthesia. Six years later, the American Medical Association did as well.

MODERN TOOTHPASTE

Washington Sheffield

COLLAPSIBLE TOOTHPASTE TUBE

Washington Sheffield and Lucius Sheffield

New London, Connecticut dentist Washington Sheffield is considered the inventor of modern toothpaste.

In the mid-1870s, New London, Connecticut dentist Washington Sheffield developed a mint-flavored dental crème to replace the tooth powder that was commonly used to clean teeth. He flavored it with extracts of mints and made batches of it to sell in his thriving dentist office, where he also performed dental surgery.

Several years later, his dentist son Lucius visited France, where he observed artists squeezing paint from collapsible tubes. When he returned home, Lucius and his father inserted the toothpaste into similar tubes, and the first toothpaste tube was created. It was also an improvement in dental hygiene—up until that time, Sheffield's dental crème was sold in porcelain jars, and it was common for all members of a household to dip their toothbrushes in the same jar.

Sheffield registered the trademark of Dr. Sheffield's Creme Angelique Dentifrice in 1881. A *New London Telegram* newspaper advertisement for that year states, "It is composed of the best ingredients known for cleaning the teeth, neutralizing the acids of the mouth, and preventing decay; and contains nothing that can in any way affect the teeth deleteriously." Other inventions of Washington Sheffield's included improved dental crowns and bridges.

In the 1890s, Washington Sheffield built a laboratory and a factory behind his home in New London. After he passed away in 1897, Sheffield's two grandsons continued the large-scale manufacturing of his original formula toothpaste, along with other manufacturers' toothpastes, throughout the United States and other countries. In 1911, after receiving a tube patent,

A Crest brand toothpaste tube. The collapsible toothpaste tube was invented by Washington Sheffield and Lucius Sheffield. *Author's collection.*

they established the New England Collapsible Tube Company, which manufactured and sold empty tin tubes for various pharmaceuticals and toiletries. Today, the successor to the Sheffield company still produces toothpaste in New London, Connecticut.

"Father of American Physiology"

William Beaumont

Lebanon, Connecticut, born and raised, William Beaumont is known for his pioneering work on the human gastrointestinal tract. The son of Congregationalist parents, Beaumont left home at age twenty-one. It was said he had with him only a horse, a small sleigh and a barrel of cider. After a stint teaching in New York State, he traveled to Vermont, where he apprenticed with a local medical doctor. In the War of 1812, he volunteered to serve as an army surgeon. After the war, Beaumont served at the fort on Mackinac Island, at the upper tip of Lake Huron.

In June 1822, Dr. Beaumont was called to attend to a young French-Canadian trapper, Alexis St. Martin, who suffered an accidental shotgun wound to his abdomen that left a cavity the "size of a man's fist" with the burnt edge of a lung sticking out from it. After treatment by Beaumont, to the surprise of all, the man recovered. Beaumont took the man into his own home, and although unable to close his wound, he cared for him there for three years. He noticed that through the gunshot hole, he could witness the process of digestion. With direct access to the gastric contents, Beaumont was able to perform hundreds of experiments over the next ten years.

In 1833, Beaumont published his findings in *Experiments and Observations on the Gastric Juice and the Physiology of Digestion.*

Becoming known as the "Father of American Physiology," Dr. Beaumont died in 1853. Although his wound never fully healed, St. Martin worked at manual labor, married, fathered children and traveled extensively. He passed away twenty-seven years after Dr. Beaufort at age seventy-eight.

Vitamin A

Lafayette B. Mendel and Thomas B. Osborne

Lafayette B. Mendel and Thomas B. Osborn collaborated together as researchers at Yale University. In 1913, they discovered a fat-soluble nutrient in butter that became known as vitamin A.

Born in 1872 in south central New York State, Lafayette B. Mendel was the son of Jewish immigrants from Germany. As an undergraduate student at Yale University in New Haven, he studied the classics, economics and the humanities. At age nineteen, he was the youngest student in his Yale graduating class. He proceeded on to postgraduate research in physiological chemistry at Yale's Sheffield Scientific School. Mendel received his PhD in 1893 at age twenty-one. After research in Germany, he returned to the Sheffield School and spent the rest of his career at Yale. He was appointed an assistant professor and in 1903 a full professor. Mendel was one of the first tenured Jewish professors at Yale University and the first to be named a Sterling Professor (1921).

Mendel's fellow researcher Thomas Osborne of the Connecticut Agricultural Experiment Station was born in New Haven in 1859 and spent his whole life in New Haven. Eli Whitney Blake, the inventor of the stone crusher, was his grandfather, and Eli Whitney, the inventor of the cotton gin, was his great-great-uncle.

Mendel and Osborne studied the effects of proteins on the development and health of rats. They observed that if young rats were given only proteins, sugars and starch, their growth was "restricted." However, if milk fat was also part of their diet, growth was not affected. In searching for the ingredient in milk fat that caused this, they discovered vitamin A in 1913.

University of Wisconsin researchers Elmer McCollum and Marguerite Davis, while also studying rats, independently discovered vitamin A at the same time. Both teams submitted their findings to *Journal of Biological Chemistry* in 1913, and both were published in the same issue. However, since McCollum and Davis made their submission twenty days before Mendel

Left: American biochemist Professor Lafayette B. Mendel was co-discoverer of vitamin A in 1913. *Library of Congress.*

Below: Professor Lafayette B. Mendel's house in New Haven, Connecticut, is a National Historic Landmark. *Author's collection.*

and Osborne, they were initially given credit. Over their careers, Mendel and Osborn collaborated on over one hundred scientific papers.

Mendel and Osborne both died in New Haven—Osborne in 1929 and Mendel in 1935. Mendel's house in New Haven is a National Historic Landmark.

"THE FATHER OF NEUROSURGERY"

Harvey Cushing

Harvey Cushing was born in 1869 and was admitted to Yale University. Medicine ran in his family—his father, grandfather and great-grandfather were medical doctors. Four years after graduating from Yale, he received his MD from Harvard Medical School. In 1896, he studied surgery at Johns Hopkins Hospital and eventually chose to specialize in neurosurgery. At the turn of the century, Cushing was one of the first medical doctors to throw his support behind Italian pediatrician Scipione Riva-Rocci's new device to measure blood pressure before, during and after an operation. Ultimately, the device turned out to be the earliest method of accurately measuring blood pressure.

In 1913, Cushing became surgeon-in-chief at Harvard. During World War I, he was a senior consultant in neurological surgery for the American Expeditionary Forces in Europe. At age sixty-three, Cushing retired from his position and moved back to New Haven as Yale Medical School's Sterling Professor of Medicine in Neurology.

Shortly after returning to Yale, Dr. Cushing moved his collection of "brain specimens, tumor specimens, microscopic slides, notes, journal excerpts and over 15,000 photographic negatives dating from the late 1800's to 1936" from Massachusetts to Yale. He had been adding to it for over thirty years. For four years, Cushing and neuropathologist Louise Eisenhardt worked to gather photographic copies of each history for which he had a pathological specimen. Today, the collection is housed in the Cushing Center, which opened in 2010.

Harvey Cushing's name is best known today because of Cushing syndrome, which is named after him. It occurs when a body has too much of the hormone cortisol over time. Too much cortisol can cause the physical signs of Cushing syndrome—high blood pressure, bone loss and type 2 diabetes. It was Cushing himself who first identified the disease. He died in New Haven in 1939.

Tissue Cultures

Ross Granville Harrison

Longtime Yale University biologist Ross Granville Harrison was the first person to develop animal-tissue cultures.

An 1889 graduate of Johns Hopkins University, Harrison earned a PhD there in 1894. In 1899, he received a doctor of medicine degree at Germany's Bonn University. At Yale from 1907 until his retirement in 1938, Harrison was a professor of comparative anatomy and biology, chairman of the zoology department and director of the Osborn Zoological Laboratory. When Harrison and his family moved to New Haven, they lived in the building on York Street that is now the home of the private club Mory's. It was only one-third of a mile from Harrison's laboratory at 2 Hillhouse Avenue.

Each year since 1822, the City of Philadelphia has given its prestigious John Scott Award to people whose inventions have contributed to the "comfort, welfare and happiness" of mankind. In 1925, Harrison won the award for "Method of tissue culture, etc."

Harrison passed away in 1959. A large marker in New Haven's Grove Street Cemetery reads, "Teacher—Scientist—Administrator. Professor of Zoology at Yale University, 1907–38. He was the discoverer of the tissue-culture method of studying the development of living organisms."

Shortly after Harrison's death, his colleague, zoology professor John Spangler Nicholas, wrote *Ross Granville Harrison: A Biographical Memoir.* Nicholas tells us a little about Harrison the man:

> *Harrison's interest in the environs of New Haven was a dual one. He first wished to locate the breeding pools of the amphibia and secondarily he enjoyed hiking. With two of his friends,* [physiologist] *Yandell Henderson and* [chairman of Yale's Department of Classics] *George Hendrickson, there was little of the terrain surrounding New Haven that was not covered. They referred to themselves as the 3H Walking Club. Harrison's zest for walking embarrassed many younger men and a short jaunt would convince them that here was a man who could walk both fast and long.*

MODERN VACCINES

John F. Enders

Often called the "Father of Modern Vaccines," John F. Enders was born in West Hartford, Connecticut, into a well-to-do family in 1897. His father was an executive at Hartford National Bank and helped author Mark Twain with financial matters. Enders later would remember that when Twain visited his home, he was always dressed in his trademark white suit.

Enders attended Hartford's Noah Webster public grammar school and between ages fifteen and eighteen attended St. Paul's boarding school in Concord, New Hampshire.

Before settling on his chosen field, Enders served as a flight instructor in the U.S. Naval Flying Corps during World War I, received a bachelor's degree in English from Yale University, worked in real estate and enrolled at Harvard in the English literature graduate program. Finally, he obtained a PhD in bacteriology and immunology at Harvard University and began one of the most successful medical research careers in history.

Enders shared the 1954 Nobel Prize in Physiology or Medicine with two colleagues "for their discovery of the ability of poliomyelitis viruses to grow in cultures of various types of tissue." Enders died in Waterford, Connecticut.

A biographical memoir written by Thomas H. Weller and Frederick C. Robbins, Enders's two associates who shared the 1954 Nobel Prize with him, was published by the National Academy of Sciences in 1991. It mentions a little about his personal life:

> *His major nonscientific interests were fishing and playing the piano. The family spent summers in the Enders' compound at Waterford, Connecticut, from which they launched powerboat outings on Long Island Sound in search of striped bass. Enders himself made a pilgrimage each summer to his brother's fishing club in New Brunswick. If they were successful, salmon packed in ice would arrive at the laboratory.*
>
> *Playing the piano was for the most part a private matter for Enders, and his interests ranged from Bach to Joplin. One exception, however, was the annual Christmas party he held at his home for his laboratory staff, and which regularly concluded with Enders at the piano, accompanying Christmas carols.*

In addition to the Nobel Prize, Enders was featured as *Time* magazine's Man of the Year in 1961, and in 1963, he was selected to receive a Presidential Medal of Freedom by President John F. Kennedy.

MELATONIN

Aaron B. Lerner

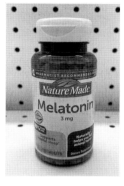

Melatonin was isolated by Aaron B. Lerner and his colleagues at Yale University in 1958. *Author's collection.*

Melatonin is a hormone secreted by the pineal gland that regulates human sleep-wake cycles. In 1958, it was isolated by American medical doctor and dermatology professor Aaron B. Lerner and his colleagues at Yale University.

Based on experiments with frogs, they thought melanocyte-stimulating hormone (MSH) might be effective in treating disorders of skin pigmentation like vitiligo. In 1956, Lerner isolated MSH at Yale and named it melatonin. It was later discovered that it is a key to the body's circadian rhythm, which is a natural process that regulates the sleep-wake cycle. Today, melatonin is used to treat insomnia and jet lag.

A resident of Woodbridge, Lerner possessed both a medical degree and a doctorate in physiological and physical chemistries. He was named the first chairman of Yale's dermatology department when he was in his late thirties. In 1973, he became the first dermatologist elected to the National Academy of Sciences. He died in New Haven in 2007 from complications of Parkinson's disease at age eighty-six.

MILK OF MAGNESIA

Charles H. Phillips

After English pharmacist Charles H. Phillips arrived in the United States, he first settled in New Jersey. In 1848, he moved to the Glenbrook section of Stamford, Connecticut, and the following year, he opened the Phillips

Milk of magnesia was invented by pharmacist Charles H. Phillips in 1873. His factory and his summer residence were in Stamford. *Author's collection.*

Camphor and Wax Co. near his home. The company's products included camphor, wax and cod liver oil.

In 1873, Phillips invented and patented milk of magnesia, which consisted of an 8 percent suspension of hydrate of magnesium and water. Its name came from the milky color of the product.

When Charles Phillips died in 1888 at age sixty-nine, his company, the C.H. Phillips Chemical Company, was headquartered in New York City, while his factory was still in Stamford. His summer residence mansion was also in Stamford.

Phillips's four sons took over control of his business. In 1923, it was purchased by the Sterling Products Corp., which promoted the product milk of magnesia above all others and introduced milk of magnesia toothpaste and tablets. The last year that milk of magnesia was made in Stamford was 1976. Bayer HealthCare purchased Sterling in 1995 and continued the production of milk of magnesia.

FIRST SUCCESSFUL CLINICAL USE OF PENICILLIN IN THE UNITED STATES

Yale New Haven Hospital

The first patient to be treated by the new antibiotic penicillin in the United States was a thirty-three-year-old mother named Ann Miller. It was 1942, and the place was New Haven Hospital (now Yale New Haven Hospital). Miller was dying of a streptococcal septicemia infection that resulted from a miscarriage. In a desperate effort to save her, doctors gave her sulfa drugs and other antibiotics, conducted a blood transfusion, performed a hysterectomy and even tried rattlesnake serum. Her temperature rose to 106.5 degrees.

Miller's doctors spoke with Dr. John Fulton, a neurophysiologist at Yale School of Medicine who studied at Oxford University with Howard Florey, a researcher who was studying penicillin. It had been discovered by Scottish physician and scientist Alexander Fleming in 1928 and given to a handful of British patients. By 1942, it still had not been approved for use by humans in the United States. Fulton sought and received that approval.

Merck Laboratories had produced a small amount of penicillin, and 5.5 grams of it was rushed by air from Washington, D.C., to New Haven and carried to the hospital by a Connecticut state trooper.

After receiving her first doze of penicillin, Anne Miller's four-week fever returned to normal in less than twenty-four hours. The streptococcal infection was defeated, and Miller lived another fifty-seven years before dying in Salisbury, Connecticut, in 1999. Her case was an important step in convincing pharmaceutical companies of the importance of mass-producing penicillin. It soon was saving the lives of servicemen and civilians who contracted streptococci, staphylococci and pneumococci during World War II. It ultimately became the best-known and most widely used antibiotic in the world.

FIRST INTENSIVE CARE UNIT FOR NEWBORN BABIES

In 1960, New Haven Hospital (later renamed Yale New Haven Hospital) introduced the world's first intensive care unit for newborn babies. Specializing in the care of ill or premature newborns, it was designed by pediatrician Louis Gluck (1924–1997), who has been called the "Father of Neonatology." At the same time as the unit was established, a perinatal service was begun to monitor fetuses for possible problems. Gluck also invented a test to determine if a fetus's lungs were mature.

Before the New Haven unit was built, premature babies were commonly isolated in small cubicles with little contact with doctors and nurses. Gluck, an expert in infectious diseases, applied his knowledge to creating a much safer infant ICU. He later went on to establish a second intensive care unit for newborn babies at the University of California at San Diego.

In 2018, Yale New Haven Children's Hospital continued the progress as it opened a sixty-eight-bed state-of-the-art neonatal intensive care unit. The two-floor unit is designed to "enhance family-centered care, provide support and improved outcomes and advance clinical research."

FETAL HEART MONITOR

Orvan W. Hess and Edward H. Hon

In the 1930s, obstetrician and gynecologist Orvan W. Hess, a resident at New Haven Hospital and later a research fellow at the Yale University medical school, began looking at the feasibility of electronically measuring fetal heart activity. At the time, the only device available to monitor a fetus's heartbeat was the stethoscope. However, it would also pick up the mother's heartbeat, and when the mother was having contractions, her heartbeat overshadowed the fetal heartbeat.

Years later, along with fellow medical doctor China-born Edward H. Hon, MD (1917–2006), Hess pioneered the development of the fetal heart monitor. (After medical school in California, Dr. Hon had completed his obstetrics and gynecology residency and fellowship at the Yale University Medical Center.)

In 1957, an article in the journal *Science* by Hess and Hon described their six-foot-high-by-two-foot-wide monitor. Joined by the head of Yale's electronics laboratory, Wasil Kitvenko, they continued to improve the machine.

Orvan W. Hess died in 2002 at Yale New Haven Hospital at age ninety-six. Edward Hon passed away four years later at age eighty-nine.

Today's fetal heart monitors are far smaller than the early Hess and Hon equipment, and over the decades, they have greatly reduced the number of stillbirths.

NATURAL CHILDBIRTH

Yale New Haven Hospital

In 1949, Yale New Haven Hospital became the first hospital in the United States to provide natural childbirth for all obstetrical patients. It was organized under the direction of Professor of Obstetrics and Gynecology Herbert Thoms, MD.

At that time, Dr. Thoms wrote of a young woman who was in her first pregnancy. Early in her labor, she expressed a wish to have a "natural childbirth." Thoms reported that she was "highly successful," and the hospital conducted sixteen similar deliveries in the subsequent six months. Under Thoms, Yale University School of Medicine initialed a study of natural childbirth, and many other institutions and individuals learned from their results.

Rooming In at Yale New Haven Hospital

In 1946, Yale New Haven Hospital became the first hospital in the United States to allow healthy newborns to stay in rooms with their mothers.

Today, the hospital's website describes the program well:

> *When you give birth at Yale New Haven Hospital, nurses will care for you and your baby together. Most of the day, your baby will remain at your bedside. This is known as "rooming in." This helps you quickly learn your baby's cues and get to understand what he or she wants. Many new moms also sleep better when their baby is in the room. Research has shown that rooming in has these benefits:*
> - *Being close to mom makes it easier for babies to get used to life outside the womb.*
> - *When babies feel their mom's warmth, hear her heart beat and smell her, they feel safe.*
> - *Babies get to know their mom by using their senses. They are able to tell the difference between their mother's smell and that of another woman by the time they are one to two days old.*
> - *Baby's attachment instinct is highest during the first days of life. Early attachment has a positive effect on baby's brain development.*
> - *Rooming in helps babies regulate their body rhythms. This includes heart rate, body temperature and sleep cycle.*
> - *In special circumstances, your baby can be brought to the nursery.*

Artificial Heart

Robert Jarvik

Robert Jarvik was born in 1946 and grew up in Stamford, Connecticut. As the son of a doctor, it is not surprising that he invented a surgical stapler. What was surprising was that he was a teenager at the time. After getting his medical degree at the University of Utah, he decided to enter the field of medical research.

In 1953, when Jarvik was seven years old, the first successful artificial heart-lung machine was used while doctors were working on a patient's heart and lungs. The doctors were able to operate while the machine took over the heart and lung functions. But it was a temporary measure, and the machine was disconnected when the operation was over. Jarvik sought a device that could function as a heart and be located inside a person's body.

At age thirty-six, Jarvik invented an artificial heart made from aluminum and polyurethane. Known as the Jarvik-7, it was designed to be permanently implanted in a heart recipient's body. The first patient was sixty-one-year-old dentist Barney Clark, who was not a candidate for a human heart transplant. After receiving the mechanical heart in December 1982 from cardiothoracic surgeon William DeVries, Clark was connected to a 350-pound air compressor and lived for 112 days. It was the first time in history that someone had lived anywhere near that long with a mechanical heart.

Robert Jarvik continued to improve his mechanical heart and through the 1980s implanted the Jarvik-7 in over seventy patients. Most notably, in 1984, the Jarvik heart was implanted in William J. Schroeder, who lived 620 days.

Today, artificial hearts of the Jarvik type are routinely used—often for extended periods—in patients who are waiting for human donors.

LYME DISEASE

Stephen E. Malawista and Allan C. Steere

In 1975, two mothers who lived in towns along the east bank of the Connecticut River encountered an unusual outbreak of arthritis in their families and communities. They asked the Connecticut State Department of Health and the Yale School of Medicine for help. Medical doctors Stephen E. Malawista, who was chief of rheumatology at the Yale School of Medicine, and Allan C. Steere, then a Yale postdoctoral student, responded, as did Doctors David R. Snydman and Francis M. Steele from the Connecticut State Department of Health.

Old Lyme, Lyme and East Haddam were investigated. Of their combined 12,999 residents, 39 children and 12 adults were diagnosed with juvenile arthritis or arthritis with an unknown cause. Although physical examinations and bloodwork did not identify a cause, about one out of four of the patients reported the appearance of a skin lesion "with an expanding bull's-eye pattern four or more weeks preceding the onset of arthritic symptoms." All of the patients lived in the wooded areas of their towns.

Steere and Malawista saw a similarity between the Connecticut patients' rashes and a European skin lesion, erythema migrans (EM). They hypothesized that a tick might be responsible. In subsequent years, with additional outbreaks, the two doctors discovered the Lyme arthritis was thirty times greater on the east side of the Connecticut River than the west side. This mirrored the deer and deer tick distributions. Scientists later determined that Lyme disease in the United States is transmitted by the deer tick (*Ixodes scapularis*).

Steere and Malawista's work led to the discovery of Lyme disease. Dr. Malawista, after a remarkable career at Yale, died in Hamden at age seventy-nine in 2013. After a career that includes more than three hundred articles on Lyme disease, Dr. Steere is today a professor of medicine at Harvard Medical School and director of translational research in rheumatology at Massachusetts General Hospital, where his lab "specializes in the identification of novel antigens, including those from infectious agents or self-proteins, the elucidation of immune responses to these antigens, and the determination of microbial and host genetic factors that underlie adverse clinical outcomes."

DISCOVERY OF THE LINK BETWEEN HIV AND AIDS

Robert Gallo

Born in Waterbury, Connecticut, Robert Gallo played a key role in establishing that Human Immunodeficiency Virus (HIV) caused Acquired Immune Deficiency Syndrome (AIDS).

Gallo grew up in the Waterbury house that his grandparents purchased when they emigrated from Italy. Gallo's father owned a welding company. Young Gallo became interested in a career in medicine when he was twelve years old. His younger sister died of leukemia, and he was impressed by the skill and compassion of the doctors who cared for her.

After a bachelor's degree in biology from Providence College in 1959, an MD from Jefferson Medical College in 1963 and a clinical clerkship at Yale University School of Medicine in 1963, Gallo completed his residency at the University of Chicago in 1965.

One of the longest-running disputes in medical history concerned the discovery of HIV's role in causing AIDS. In the early 1980s, questions were raised on who discovered the link between HIV and AIDS first. The two possibilities were Robert Gallo's laboratory at the United States National Cancer Institute and Luc Montagnier's laboratory in France at the Pasteur

Institute. Although in 2008 the Nobel Prize for Physiology or Medicine was awarded only to Luc Montagnier and his colleague Françoise Barré-Sinoussi, the controversy continues.

In 1986, Gallo shared the highly respected Albert Lasker Award for "HIV as the cause of AIDS" with Luc Montagnier and Myron Essex. Gallo previously (1982) had received an Albert Lasker Award for "Oncogenes that transform normal cells into cancer cells."

In later years, Gallo led efforts to develop a blood test for HIV, which enabled a screening for the AIDS virus. This allowed patients to be diagnosed earlier with AIDS and reduced the number of people contracting it from blood transfusions.

For thirty years, Gallo worked at the National Institutes of Health's National Cancer Institute, where for twenty-three years he directed its Laboratory of Tumor Cell Biology. In 1995, he co-founded and became director of the Institute of Human Virology at the University of Maryland. In recent years, Gallo and his team have been working on developing a vaccine for AIDS. Gallo also co-founded the Global Virus Network "to position the world to rapidly respond to new or re-emerging viruses that threaten mankind, to achieve collaboration among the world's leading virologists, and to support next-generation training."

Over the decades, Gallo has been awarded thirty-five honorary doctorates, has had more than 1,200 scientific articles published and was inducted into the National Inventors Hall of Fame in 2004.

MOBILE GENETIC ELEMENTS

Barbara McClintock

Hartford-born Barbara McClintock won the 1983 Nobel Prize in Physiology or Medicine for "her discovery of mobile genetic elements." According to her Nobel Prize biography, she "proved that genetic elements can sometimes change position on a chromosome and that this causes nearby genes to become active or inactive." The biography goes on to remark, "She was shy and anything but a careerist, but at the same time she also realized the importance of what she had achieved, not least of all in her role as an example for other women."

Raised in a lower-middle-class family, she entered Cornell's College of Agriculture in 1919. It was there that she became interested in cytogenetics,

Hartford-born Nobel Prize winner Barbara McClintock in her laboratory in 1947. *Smithsonian Institution.*

which is the branch of genetics that deals with the study of the structure and function of chromosomes. After earning her bachelor's degree in 1923, she stayed at Cornell University to receive her master's in 1925 and her doctorate in 1927. After receiving her doctorate, McClintock engaged in her research interests full time at Cornell and later at the University of Missouri and the Cold Spring Harbor Laboratory in New York. Her work included tracing genes through generations of corn plants. Her discoveries related to how genes control the growth and development of cells.

Thirteen years before receiving the Nobel Prize, Barbara McClintock was given the National Medal of Science in 1970. It is the "highest honor bestowed on American scientists and engineers by the president of the United States." The National Science Foundation website states:

> *By the early 1970s, innovations in molecular biology and genetic engineering began to confirm McClintock's earlier theories. These new findings sparked a renewed interest in, and appreciation for, her work. In addition to receiving the National Medal of Science from President Richard M. Nixon—the first woman honored with the medal—McClintock was the first recipient of a MacArthur Foundation grant and, in 1983, she became the first woman to win an unshared Nobel Prize.*

Barbara McClintock passed away at age ninety in 1992. Dr. James A. Shapiro, professor of biochemistry and molecular biology at the University of Chicago, called her "the most important figure there is in biology."

CATALYTIC PROPERTIES OF RNA

Sidney Altman

Longtime Yale University professor Sidney Altman (b. 1939), along with University of Colorado's Thomas R. Cech (b. 1947), won the 1989 Nobel Prize in Chemistry for the discovery of catalytic properties of RNA.

As stated in the Nobel Prize organization's description, "Enzymes are substances that speed up the chemical processes in organisms' cells without being consumed. It was long thought that all enzymes were proteins. Sidney Altman and Thomas Cech demonstrated that RNA can also function as an enzyme."

Canadian-born Sidney Altman was the son of a father and mother who were respectively a grocery store owner and textile millworker. Altman writes in his Nobel Prize autobiography, "No sacrifice was too great to forward our education and, fortunately, books and the tradition of study were not unknown in our family." After studying physics at the Massachusetts Institute of Technology and Columbia University, Altman became a full-time member of Yale University's faculty in 1971.

Today, Altman is professor emeritus of molecular, cellular and developmental biology at Yale.

Chapter 4

TRANSPORTATION

Few American states can match Connecticut's reputation as the origin of significant advances in virtually every mode of transportation. From the development of the first military submarine during the American Revolutionary War, which was invented by a Connecticut man, to the world's first nuclear-powered submarine in the mid-twentieth century, which was built in Groton, the state has pioneered sea power. From the railroad tracks that early in the 1800s crisscrossed the state to the world-class superhighways of the mid-1900s, Connecticut has been an innovator. The world of air travel is no exception, with Connecticut's invention of the first practical helicopter and the location of the company that designed and manufactured the propellers on three-quarters of the U.S. military planes of World War II. In the late nineteenth century, Connecticut people also pioneered some of the first automobiles in the world. Over one hundred years before the advent of the modern electric and hybrid cars, it was a leader in the invention, development and manufacture of electric automobiles.

STEAM-POWERED BOAT

John Fitch

Industrialist John Fitch was born to a Windsor, Connecticut farming family in 1743. Early in life, he was employed as a clockmaker's apprentice, a brass foundry operator and a silversmith. After working as a supplier to

John Fitch's steamboat of 1790 being used to transport passengers. *From the 1900 edition of* Appleton's Cyclopædia of American Biography.

the Continental army during the American Revolutionary War, he became absorbed with creating a steam-powered boat. Although Benjamin Franklin complimented Fitch on his invention and George Washington received him as a guest at his Mount Vernon home, the Continental Congress denied him financial aid for his project. Eventually, four states—New York, New Jersey, Pennsylvania and Delaware—granted him the exclusive rights to operate his steamboats on their waters.

The first successful trial of a steamboat was made by Fitch as he navigated his forty-five-foot-long craft on August 22, 1787, on the Delaware River. Movement was provided by six steam-powered oars on each side of the boat powered by a continuous chain. He proceeded to build and put into operation a larger steamboat that carried passengers and freight between Philadelphia and New Jersey. A 1790 Pennsylvania newspaper advertisement stated, "The steamboat is now ready to take passengers every Monday, Wednesday, and Friday, and to return on Tuesday, Thursday, and Saturday."

In 1791, Fitch was awarded a U.S. patent for a steamboat, but he didn't succeed commercially with the boats, which traveled at about eight miles per hour. He also obtained a French patent and traveled to France on an unsuccessful mission to strike a deal with the French government.

Fitch was far more an inventor than a businessman and never saw much success with his steamboats. He incurred financial loses and didn't obtain the support of investors or the interest of the general public to an extent that would allow his invention to achieve its potential. This left the way open for Robert Fulton to develop his successful steamboat years after Fitch died. Fitch predicted this when he said, "The day will come when some more powerful man will get fame and riches from my invention; but nobody will believe that poor John Fitch can do anything worthy of attention."

In 1817, a committee of the New York legislature was charged with deciding the identity of the inventor of the steamboat. After examining all available documents on the Fulton and Fitch vessels, it concluded, "The steamboats built by Livingston [an American statesman and business partner of Fulton] and Fulton are in substance the invention patented to John Fitch in 1791."

EARLY AUTOMOBILES

Hiram Percy Maxim

The son of Sir Hiram Stevens Maxim, who invented the portable machine gun, Hiram Percy Maxim's life can be summarized by a long listing of his inventions. One of the most significant was a gasoline-powered tricycle, which he built by 1895.

In 1897, Maxim was awarded a patent on an electric car. In the patent application, Maxim stated that he "invented certain new and useful Improvements in Motor-Vehicles" and that some "relate especially to the means of propulsion, the source of power being an electric storage battery carried by the vehicle; but other features of the invention might be applied to vehicles propelled by other motive power. I have improved the construction of the running-gear to the end that it may be very strong and well adapted to withstand rough usage without Being exceedingly heavy."

In 1899, Maxim moved to Hartford and lived there until his death in 1936. The year 1899 was when the Columbia Automobile Company used Maxim's invention to build a gasoline-powered four-wheeled vehicle.

In 1904, Maxim, along with Harry M. Pope and Herbert W. Alden, applied for a U.S. patent on an electric car. They stated in their application:

One purpose which we have had prominently in view has been the production of apparatus which can be operated, managed, and cared for by inexperienced and even unintelligent persons without danger of injury to themselves or to others or to the apparatus or to the vehicle to which it may be applied, and we have also sought to overcome as far as possible all of the defects in other apparatus of like general nature so far as they have become known to us.

Efforts by Maxim to improve the gasoline-powered automobile led to other inventions, such as the exhaust muffler. In addition to motor vehicles, Maxim was involved in many other projects. He was an early aircraft enthusiast, flying what was one of the first planes ever flown over Hartford. He also helped found the Aero Club of Hartford. For many years, he was chairman of Hartford's Aviation Commission, and he was responsible for the establishment of Hartford's airport, Brainard Field.

In the early days of radio, Maxim was a pioneer. In 1914, he founded the amateur American Radio Relay League (ARRL), which today is named the National Association for Amateur Radio, with its national headquarters in Newington, Connecticut. That station is named the Hiram Percy Maxim Memorial Station.

A 1932 article in *Life* magazine mentioned that Maxim's office was in Hartford, as was his city home, which sat on North Whitney Street across from Elizabeth Park. In addition, he had a home at the mouth of the Connecticut River, where he had his thirty-eight-foot motor yacht, *Moby Dick*. The article also revealed an interesting fact about Maxim and his inventive process: "The Maxim inventions in experimental stages are developed in a medium of tin. In his laboratory, Maxim molds, pounds, twists and cuts this household material like a boy let loose in a barn."

BICYCLES

Pope Manufacturing Co. of Hartford

After service in the Union army in the American Civil War, Albert Augustus Pope became interested in the future of bicycles. In 1878, he arranged for the Weed Sewing Machine Co. in Hartford to manufacture them for him under the brand Columbia. As sales rose, Pope bought Weed's factory. The first bicycle to be made in Hartford—the High Wheeler—was released in 1878. In 1892, Pope Manufacturing offered for sale the Columbia Century, which was the first bicycle with pneumatic tires to be produced by Columbia.

Pope advertised its Hartford Columbia Bicycle Factory as "the largest and most expensive factory in the world devoted exclusively to the manufacture of bicycles."

In 1894, Pope consolidated its main offices in Hartford, and in 1898, it introduced bicycles with wheels of equal sizes. Additionally, making the bicycles with hollow tubing made them far lighter and easier to be used by children and smaller adults. The drop frame and skirt guards on the chains and rear wheels allowed them to be used by women who wore the fashions of the day. In 1896, women's suffrage leader Susan B. Anthony said, "Let me tell you what I think of bicycling. I think it has done more to emancipate women than anything else in the world. I stand and rejoice every time I see a woman ride by on a wheel."

Four years later, Columbia began production of its first chainless shaft drive bicycle. Between 1905 and 1913, the Pope Manufacturing Co. consolidated its manufacturing to the plant in Westfield, Massachusetts, while its main offices were retained in Hartford, Connecticut.

Company founder Colonel Albert Augustus Pope died in 1909. In 1915, the Pope Manufacturing Co. was reorganized after bankruptcy and renamed the Westfield Manufacturing Company. In 1933, it was acquired by the Torrington Company of Torrington, Connecticut. It added steel school furniture to its bicycle product line in 1952 to keep its production facilities operating during the bicycle's off-season. Both product lines used some of the same machinery. In 1950, Columbia released the News Boy Special, which was a precursor to today's mountain bikes.

In 1961, the company was renamed the Columbia Manufacturing Company. After declaring bankruptcy again in the early 1990s, it continued on with the production of bicycles and school furniture. In the 2000s, the company rereleased its 1937 and 1952 models.

ELECTRIC CARS

Albert Augustus Pope

In the 1890s, Pope was manufacturing about one-quarter of a million bicycles in his Hartford factory. He also added an automobile division and developed and manufactured the first commercial electric automobiles.

A year after establishing the Columbia Electric Vehicle Co., Pope introduced and began selling the first electric-powered cars. In the year

1899, he built hundreds of cars in Hartford. Many were electric powered. For a few years, Pope was the top automobile manufacturer in the country.

In 1900, about one-third of the automobiles in the United States were electric powered. America's most famous inventor, Thomas Edison, preferred electric vehicles, and so he prioritized the development of electric vehicle batteries. A 1901 ad for a Columbia electric car noted its sixth year in business and touted a "radius on each battery charge" of forty miles and a maximum speed of fourteen miles per hour.

By 1905, Pope was making fifteen models of electric cars. An advertisement that year for its Model 36 Pope Waverley electric car stressed its ease of operation: "Simply throw forward controller and steer. No noise, no odor, no jar, jolt or vibration….Jump in, throw the lever, and whiz away without a second's delay!" Priced at $900 (with an optional leather top for an additional $50), the ad claimed it was the most popular electric car in the world.

However, in the first years of the twentieth century, oil was discovered in Texas and Henry Ford introduced his inexpensive gas-powered cars. The competition was too much for Pope and his company. Sales and profits declined, and Albert Augustus Pope passed away in 1909 at age sixty-six.

Today, Pope Park in Hartford keeps the inventor's name alive. It was donated to the city by Pope in 1895 at the height of his company's success. The twenty-first-century success of electric-powered vehicles like Teslas and Chevrolet Bolts has shown Pope to be ahead of his time—by over a century.

ELECTRONIC STREET CARS

Frank Julian Sprague

Inventor Frank J. Sprague was born in Milford, Connecticut, in 1857. When he was seven, his mother died and his father, the superintendent of a hat factory, sent the boy to live with an aunt in Massachusetts. Years later, Frank mentioned how he sold lemonade and apples to raise money for the family.

During his career, Frank Sprague is credited with developing reliable electric motors that led to practical trolley cars, today's subway trains and electric elevators. The latter could more efficiently handle heavier loads than the former non-electric ones. His work on subways and elevators was particularly important for the growth of large cities. In the early twentieth century, it was said that what Thomas Edison did for the development of electric light, Frank Sprague did for electric power.

As a cadet at the United States Naval Academy at Annapolis, Maryland, Sprague excelled in the natural sciences and mathematics, and after graduation, he was commissioned as an ensign in the United States Navy. He served from 1878 until he resigned his commission in 1883 in order to devote himself to experiments on electrical devices. He spent a short time as a scientist at Thomas Edison's laboratory in Menlo Park, New Jersey. There, he proposed "mathematical procedures and methods for increasing accuracy" that greatly benefited the laboratory.

After leaving Edison, Sprague started the Sprague Electric Railway and Motor Company. The new company relied on two of Sprague's most important inventions: the first motor that could maintain a constant speed under varying loads, and regenerative braking, which converts energy from braking into electricity and then stores it in a vehicle's battery.

Electric locomotives had been tested in the 1880s, but they were not powerful enough to pull enough cars to justify a switch from gas to electric power. Sprague's system, however, entailed having separate motors in each car with only one operator—a motorman—who controlled all of the cars from his position in one main train car.

When Sprague tried to interest investors in his system, he ran into problems. Finally, railroad developer Jay Gould, who was one of the wealthiest people in America, agreed to view a demonstration of Sprague's invention. If he was convinced, it would probably mean that Manhattan's elevated railway lines could be converted from steam power to electric power. Unfortunately, as soon as Sprague started up the device, sparks flew near Gould and the train lunged forward. Gould, both terrified and angry, immediately left.

In 1888, Sprague installed the world's first citywide electric railway system in Richmond, Virginia. The twelve-mile-long streetcar system included electric power, overhead power lines and regenerative braking. In a couple of years, there were over one hundred electric railway systems in the United States that used Sprague's electric system to replace horse-drawn cars and cable car lines. Florence, Italy, and Halle, Germany, were among the first foreign cities to adopt Sprague's system.

The United States Patent Office's annual report that was released in 1898 noted:

> *The use of electricity for power purposes has found its most notable development in the electric railway....In 1880, of the 2,050 road miles of street railway in the United States, nearly all used animal power.... The total mileage of electric railways in the United States up to October of*

1897 was 13,765 miles out of a total mileage of 15,718, of which but 947 miles were horse car lines.

In 1892, Sprague invented the first automatic electric elevator. In 1897, he created a multiple-unit electric railroad system that allowed an "unlimited numbers of motorized cars that are controlled by a master switch."

At the oldest operating trolley museum in the United States, the Shore Line Trolley Museum in East Haven, Connecticut, an exhibit titled "Frank J. Sprague: Inventor, Scientist, Engineer" was installed in 1999 to honor the inventor. Sprague was inducted into the National Inventors Hall of Fame in 2006. Today, with the reduced emissions, popularity and potential of electric vehicles, the inventions of Connecticut native Frank Sprague deserve to be better known.

One reason that Sprague is relatively unknown today is that he sold the products of his most famous endeavors. These include his Sprague Electric Railway & Motor Co., which was purchased by Thomas Edison, and his electric elevator company that was sold to Otis Elevator Co.

Late in life, Sprague moved back to Connecticut, establishing a home in Sharon. He died in 1934 and is buried in Arlington National Cemetery.

TRAFFIC LAWS

William Phelps Eno

Although little known today, longtime Westport, Connecticut resident William Phelps Eno's influence on modern life is immeasurable. Many of the traffic laws that we take for granted were devised by Eno, beginning in the days of the horse and buggy. As Eno stated in his 1909 book *Street Traffic Regulation*, "Regulation of street traffic was unknown in New York up to January, 1900....Many unnecessary hours and often the greater part of a day and night were consumed in transporting merchandise from point to point, especially in the downtown shopping districts." In the year his book was published, the New York City Police Department adopted his traffic regulations. Eno also designed Columbus Circle near New York City's Central Park

In addition to New York City, Eno created the first traffic plans for London, England, and Paris, France, including the latter's traffic circle around the Arc de Triomphe. He also had a hand in introducing the following:

- Right-hand driving
- Systems of shared intersections
- One-way traffic circles
- One-way streets
- Pedestrian safety islands
- Traffic lights
- Traffic signs
- Pavement marking
- Regulations against jaywalking
- Taxi stands
- Off-street parking
- Driver's licenses
- Vehicle registration
- Traffic tickets

Today, Eno's name lives on in the Eno Center for Transportation. This is a Washington, D.C.–based nonpartisan think tank that tackles transportation issues. Eno founded it in 1921. He passed away in Norwalk, Connecticut, in 1945.

Bicycle Bell

Bevin Bells and Frederick L. Johnson

Usually mounted on handlebars, the bicycle bell is designed to warn pedestrians, roller skaters and other bicyclists that they are being passed. In 1887, British inventor John Richard Dedicoat became the first person to patent a bicycle bell. However, many people believe that the first bicycle bell was created over twenty years earlier—in 1865—by East Hampton's Bevin Bells.

Other bicycle warning devices were devised in Connecticut during the late nineteenth century. Frederick L. Johnson of Wallingford received a U.S. patent for what he called "a new improvement in bicycle alarm-whistles" in 1885. The patent specification states that it is "a device constructed for attachment to bicycles, and by which an alarm or signal may be given by the rider to indicate his presence." In 1889, Johnson obtained another patent for a bicycle whistle.

DIRIGIBLES

Charles F. Ritchel

Dirigible pioneer Charles F. Ritchel of Bridgeport in a circa 1870 carte de visite.

Unlike earlier balloons that were first flown by French inventors in the late eighteenth century, dirigibles included propulsion and steering systems that allowed them to be directed on a course.

In 1878, prolific inventor Charles F. Ritchel of Bridgeport was awarded a patent for his one-person Ritchel Flying Machine, which was composed of a twelve-foot-diameter by twenty-four-foot-long barrel-shaped gas bag filled with hydrogen gas. A car holding a platform for an operator was suspended from it by cords and rods. The patent application stated it included "a propeller-wheel upon the front end of the machine, whereby the machine can be made to move either backward or forward, or turn to the right or left, thus enabling the operator to move the machine in any direction at will."

In May 1878, the crank-operated dirigible was flown inside an exhibition hall in Philadelphia. Because Ritchel was too heavy for his aircraft, he found a lighter person to be the operator. Then on June 12, 1878, at a ballfield behind the Colt Armory in Hartford, Ritchel held the first outdoor exhibition of his balloon with "lightweight" Mark Quinlan as its operator. The local newspaper, the *Hartford Daily Courant*, announced the price of admission as fifteen cents for adults, ten cents for children and twenty-five cents for reserved seats. Unlike the earlier flight in Philadelphia, this time the airship faced possible problems that could be caused by the wind.

This was the first controlled flight of a dirigible in North America. It was also the first dirigible flight that began and ended at the same place—leaving the ballfield, crossing the Connecticut River and returning. After the Hartford show, Ritchel toured various states with his dirigible but was only able to sell five of these "Flying Machines."

In 1911, Connecticut governor Simeon Baldwin recommended in his message to the state legislature that laws be passed controlling airships. A bill was drafted that would allow aviators to fly in the air over land or water owned by them or the land or water owned by others for which they had written permission. Anyone flying over other property would be considered trespassers unless they were licensed to do so by state authorities. The bill provided that operators must be over twenty-one years of age, pass an examination and pay two dollars for an operating license and another five dollars for vehicle registration. Also, each flying machine was required to display registration numbers that were at least three feet high.

The Connecticut aviation law passed and became the United States' first aviation law on June 8, 1911. That was a little over four months after Charles Ritchel died in his native Bridgeport. While Ritchel's idea for large transatlantic dirigibles powered by hand cranks went unfulfilled, other inventions were successful. Of his more than 150 other patents, possibly the most popular was Ritchel's funhouse distorting mirror, which is the ancestor of the mirrors still found at carnivals and fairs.

HELICOPTER

Igor Sikorsky

In 1929, Igor I. Sikorsky moved his aircraft company from western Long Island, New York, to Stratford, Connecticut. A native of Kiev in the Ukraine, he would become the most important person in the history of helicopters in the United States.

Following a childhood interest in the works of Leonardo da Vinci and the novels of Jules Verne, Sikorsky dreamed of working on a practical helicopter. In Russia, he had designed and built a multi-engine plane and created a bomber version that was used by the Russians in World War I. Escaping the country when the Communists came to power, Sikorsky immigrated to the United States in 1919. Years later, he remembered, "I was inspired by the achievements of such men as Edison, Ford, and others, and in my case particularly, the Wright Brothers." In 1923, Sikorsky founded the Sikorsky Aero Engineering Corporation in Long Island.

However, the technology would not keep up with Sikorsky's imagination until he was almost fifty years old. Finally, in 1939, his VS-300 made its first fight—of ten seconds—and became the first working helicopter with one main rotor and one tail rotor. Sikorsky himself was the pilot on that flight and afterward made it a practice to personally operate every new helicopter

The Sikorsky HH-52 Seaguard was the U.S. Coast Guard's main search-and-rescue aircraft from 1963 until the 1980s. *United States Coast Guard.*

on its first trial flight. Two years later, the VS-300 set a world record by staying aloft for one hour and thirty-two minutes.

The VS-300 led to Sikorsky's R-4, which was the first mass-produced helicopter. It also became the only Allied helicopter to serve in World War II. Soon after the war ended, Sikorsky again made history when a U.S. Army R-5 Sikorsky helicopter completed the first civilian helicopter rescue operation. On November 29, 1945, it left the Sikorsky factory in Stratford, Connecticut, to perform a rescue on Long Island Sound. An oil barge had been beached on a reef by hurricane-force winds, and it was breaking up with two men aboard. The helicopter successfully rescued the men. It was piloted by Igor Sikorsky's nephew Jimmy Viner.

During the Korean War, Sikorsky's helicopters more than proved their worth by transporting troops and rescuing injured soldiers. One of Igor Sikorsky's most famous quotations was, "If a man is in need of rescue, an airplane can come in and throw flowers on him, and that's just about all. But a direct lift aircraft could come in and save his life."

Another milestone was in 1952, when two Sikorsky helicopters made the first nonstop helicopter flight across the Atlantic Ocean. After his retirement at age sixty-eight, Sikorsky worked at the Sikorsky plant in Stratford as an engineering consultant. He passed away in 1972 at his home in Easton, Connecticut, at age eighty-three. He had seen more than five thousand of his helicopters built at the Stratford facility.

AIRCRAFT PROPELLERS

Erle Martin

Erle Martin began his career as a project engineer at United Technology's Hamilton Standard Division in 1931 and was promoted in succession to chief engineer in 1935, engineering manager in 1940 and general manager in 1946. During World War II, Hamilton Standard manufactured 75 percent of the propellers on combat planes and transports. Over the course of his career, but mostly in the 1930s and 1940s, Martin is credited with inventing more than seventy improvements to the performance and safety of aircraft propellers.

As United Technology's chief technology officer from 1960 to 1972, Martin helped move the company into jet planes and spacecraft. He rose to become vice chairman of United Technology in 1968. He died at his home in West Hartford in 1981 at age seventy-four.

AIR-COOLED AIRCRAFT ENGINES

Frederick Brant Rentschler

Frederick Rentschler helped found the Wright Aeronautical Corp. in 1909. Sixteen years later, he and aircraft engineer George J. Mead founded the Pratt and Whitney Aircraft Company. Rentschler had developed air-cooled aircraft engines to replace the liquid-cooled ones then in use. In a hearing of the Temporary National Economic Committee of the U.S. Congress in 1939, Rentschler remembered:

> *We went to Hartford and began operations on August 1, 1925, and within a few days had built up our personnel to 30 people and were*

occupying a few hundred square feet of floor space in a large empty plant of 4 stories, most of which was being used as a warehouse for tobacco. Others in the industry knew as much about Navy requirements for a 400-horsepower air-cooled engine as ourselves and it was now a matter of who first could produce a successful job. In March 1926, our engine of 410 horsepower, which we called the Wasp, successfully passed its Navy endurance test, and shortly thereafter orders were placed for 6 experimental engines of this type.

Pratt and Whitney's main task was to "develop the higher horsepower 'Wasp' radial engine for military, commercial and private aircraft." Years later, Pratt and Whitney merged with Sikorsky to become United Aircraft Corp., and in 1975, it became United Technologies.

During World War II, Pratt and Whitney Aircraft Company's Engine Division, which was headed up by Frederick Rentschler, produced Pratt and Whitney engines, Hamilton propellers, Sikorsky helicopters and Chance Vought Corsair fighters. Almost 80 percent of the air-cooled aircraft engines used by the Allied forces during World War II were manufactured by companies Rentschler founded (Wright and Pratt and Whitney) or their licensees. Pratt and Whitney alone hired as many as seventy-five thousand employees during the war.

The U.S. Air Force's Boeing B-52 Stratofortress was considered "the backbone of the Strategic Air Command during the Cold War." Powered by the J-57 jet engine, which Rentschler's team had developed, it produced twice the thrust of any other jet engine.

Rentschler was given an honorary doctor of laws degree from Hartford's Trinity College, which stated, "For helping America to gain air leadership and keep it when jet propulsion revolutionized air power."

When Frederick Rentschler passed away in 1956, he had been chairman of the board of United Aircraft Corporation for over twenty years. A *New York Times* obituary notice remarked, "This nation's air superiority is due in no small measure to Mr. Rentschler's vision and talents."

Rentschler Field, an airport in East Hartford, Connecticut, was named after Frederick Rentschler. After it closed in 1999, the forty-thousand-seat Pratt and Whitney Stadium, the home field of the University of Connecticut football team, was built on the site.

Vertaplane

Gerard P. Herrick

Ridgefield, Connecticut resident Gerard P. Herrick created an aircraft that combined the features of a traditional airplane with those of a helicopter. Begun in the 1920s, the HV-2A Vertaplane appeared at a trial flight in Philadelphia in 1937. A *Christian Science Monitor* article at the time noted that it "took off in the manner of any ordinary heavy biplane, at a speed of 180 miles an hour. At a height of 1500 feet the upper wing suddenly began to behave in a surprising manner. It flailed around with increasing speed. The ship came to a sudden stop and began an 'elevator fall' landing…which ended comparatively gently on the field." Herrick hailed his invention as "the missing link to flying safety."

The main reason the Vertaplane was never commercially successful was due to its weight; it needed to carry the combined weight of both its rotor components and its fixed wing structure. Herrick and his wife, Lois, used their Ridgefield house as a summer retreat, later moving there permanently. He died in 1955, and his wife lived at their home until her death in 1980.

Seaplanes

Edson Fessenden Gallaudet

Born in Washington, D.C., aviation pioneer and businessman Edson Gallaudet was the son of the founder of Gallaudet College for the Deaf (later renamed the American School for the Deaf).

Graduating from Yale University in 1893, Gallaudet had been on the varsity crew teams for two years. In 1896, he earned a doctorate in electrical engineering at Johns Hopkins University, and in 1898, he built and flew a glider. In 1899, he returned to Yale to the positions of physics instructor and crew coach, remaining there for two years.

In 1911, Gallaudet learned to fly an airplane at Garden City, Long Island, and obtained United States pilot certificate number 32. Even though he was in a serious airplane accident at Hempstead Plain Aircraft Field on Long Island in 1912, he continued to develop aircraft.

Edson Gallaudet started the Gallaudet Aircraft Corporation during the First World War. The company designed both military and commercial aircraft: fighters, bombers, reconnaissance planes, seaplanes, biplanes, monoplanes, triplanes, passenger planes and so on. Customers included the U.S. Navy and the Army Air Service.

The 1921 Aircraft Year Book gives a summary of Gallaudet's company's activities during the year:

> *Work upon a new series of Government Contracts, consisting of two types of seaplanes and the remodeling of a large number of DeHaviland 4's, a further development and improving of the "Chummy Flyabout" and the development and manufacture of the Gallaudet C.-3 or "Liberty Tourist" for Commercial Flying. A notable event of the year at this plant was the preparation of the planes to be used on the trip from New York to Nome, Alaska, and return by the U.S. Air Service.*

The yearbook continues with Gallaudet's Liberty Tourist:

> *In the development of the C.-3 or "Liberty Tourist" a five-seater Liberty motored biplane, the Engineering Department produced a heavier-than-air machine that is the last word in travel comfort, stability and beauty of outline. Besides a number of novel features, including fire protection, luggage receptacles and convertible cockpit, the "Liberty Tourist" had the following general characteristics:*

> *Weight, including passenger load (5) and equipment: 4675 lbs.*
> *Spread: 44 feet*
> *Length overall: 49 feet 5 inches*
> *Chord: 8 feet*
> *Gap: 6 feet*
> *Height: 10 feet 3 inches*
> *Dihedral: $1\frac{1}{2}°$*
> *Angle of incidence: 2°*
> *Wing contour R.A.F: 15*
> *Wing area inc. ailerons: 548 sq. ft*
> *Pay load: 1000 lbs.*
> *H.P. loading: 11.1 lbs.*
> *Wing loading: 8.5 per sq. ft.*
> *Cruising radius: 500 miles*
> *Estimated ceiling: 18,000 ft.*
> *Landing speed: 40 m.p.h.*
> *Maximum speed: 115 m.p.h.*
> *Power Plant, 12 cylinder 400 h.p. Liberty Motor.*

Chapter 5

BUSINESS

Connecticut inventions include items found in virtually every office: paper clips, typewriters, calculators, paper bags. What would office buildings be like if Connecticut inventors did not develop elevators or fire sprinkler systems?

PAPER CLIP

William D. Middlebrook

A paper clip is often defined as a flat, or nearly flat, piece of metal that slides over an edge of a group of two or more pieces of paper to hold them together. It does this without damaging the paper.

Born in Newtown, Connecticut, in 1846, Waterbury resident William D. Middlebrook invented the type of paper clip that is most often seen today. He also invented the machine that would produce it and was awarded a U.S. patent for this machine in 1899. On his patent application, he stated that his invention automatically manufactured "wire clips for binding or securing papers in lieu of pins." Middlebrook sold the patent to Cushman & Denison, which trademarked the name GEM for its clip. Today, that name is still found on boxes of this type of clip.

CALCULATORS

Gilbert W. Chapin

Born in Enfield, Connecticut, Gilbert Chapin worked on the family farm until he moved to New York City at age eighteen. There he worked for seventeen years in the shoe business. He also acquired experience in the newspaper and insurance businesses.

In 1889, Chapin moved to a home on Farmington Avenue in Hartford (which was/is across the street from the homes of writers Mark Twain and Harriet Beecher Stowe). He had a position as an actuary at the Society for Savings, the state's first mutual savings bank. He became a prominent member of Hartford society and later established a real estate company.

Chapin held a number of U.S. patents, but most prominent were his three patents for calculating devices. Years of business experience had shown him the deficiencies of the current equipment, and he had devised ways to improve them.

For the 1870 patent applications, he gave his address as Brooklyn, New York, while on the application for the 1900 invention he identified Hartford, Connecticut, as his home.

The three patents for calculating machines were:

1. U.S. patent 99,533 (key-operated adding device), which he described it as a "new and improved Adding-Machine."
2. U.S. patent 106,999 from 1870, for a small adder with manually rotating numeral wheels, which Chapin noted was a "new and useful Improvements in Adding-Machines."
3. U.S. patent 646,599 from 1900, for a key-operated adding device with "new and useful improvements in Adding-Machines"

In 1932, Gilbert Warren Chapin died in Hartford at age eighty-four, and he was buried in Mansfield Center Cemetery near his wife, Delia.

Portable Typewriters

George C. Blickensderfer

Born and raised in Erie, Ohio, George C. Blickensderfer began a career in dry goods. As early as age ten, he showed an interest in inventing; his first project was the development of a flying machine. This was over forty years before the Wright brothers' success in North Carolina. In the late 1880s, he and his wife moved to Stamford, Connecticut, where they lived for the rest of their lives.

In a small shop behind his house on Bedford Street, which was near the present site of the Ferguson Library, he invented his first typewriter. In 1889, he founded the Blickensderfer Typewriter Company. It began business in a Garden Street factory, manufacturing and selling his machines. In the late 1890s, he built a new factory on Atlantic Street. Later, he invented an electric typewriter but never marketed it.

In 1892, he received a U.S. patent for what many believe was the first truly portable typewriter. It included a revolving type-wheel, which was similar to IBM's type-ball in many late twentieth-century typewriters. Blickensderfer's type-wheel meant that each machine would be made with only 10 percent of the moving parts of similar machines (250 versus 2,500). This resulted in it weighing one-quarter less. Its revolutionary feature of revolving type soon attracted customers. His typewriter could be ordered with different type styles and in various foreign languages.

When World War I broke out, Blickensderfer's company, which relied on foreign sales, was hard hit. To keep his factory running, he developed a device that loaded cartridges into machine guns. His biggest customer was the French government.

Blickensderfer passed away in Stamford in 1917, and three years later, the L.R. Roberts Typewriter Company acquired his company. The twentieth century saw Connecticut become one of the typewriter manufacturing capitals of the nation. Middletown had the Noiseless Typewriter Company, which, as its name says, sold quieter machines. It later merged with another company and formed the Remington Noiseless Typewriter Corporation. Hartford had the Underwood Typewriter Company beginning in 1901. And several years later, there was the Royal Typewriter Company.

In 1913, although Connecticut had only about 10 percent the country's typewriter manufacturing companies, it employed over one-third of the industry's workers.

ELEVATORS

Elisha Otis

The Otis Worldwide Corporation is the fifth-largest corporation in Connecticut. Based in Farmington, it is also the world's largest manufacturer of elevators. Its additional products include escalators and moving walkways.

In 1852, master mechanic Elisha Otis invented the safety elevator, which used a device to automatically stop an elevator if the hoisting ropes or chains failed. The following year, Otis founded his elevator company. In 1854, demand for the new machines rose after Otis held a demonstration at the Crystal Palace Convention in New York City. In a show worthy of P.T. Barnum or Houdini, he rode an elevator high off the ground and had the rope cut. He survived, and so did his business.

In 1857, Otis installed the first passenger safety elevator at the E.V. Haughwout & Co. store in New York City. After Elisha Otis died, his company was inherited by his sons, Charles and Norton Otis, who established Otis

A 1928 Otis elevator. Seventy-six years earlier, Elisha Otis invented the safety elevator. *Library of Congress.*

Brothers & Co. The new company introduced electric elevators and became the Otis Elevator Company in 1898.

Elevators made possible the construction of today's ubiquitous skyscrapers and changed the way people lived. Before Otis's invention, the cheapest apartments were on the upper floors because of the drudgery of walking up multiple flights of steps. After the introduction of elevators, those apartments, far from the noise of the streets and sidewalks below, became the most expensive. Over the years, business executives considered it a prestigious perk to earn upper-floor panoramic views of the city, corner windowed offices or penthouse suites with their own private elevators.

One of the first private residences in the United States to have an elevator was a tower home built in Simsbury, Connecticut. The 165-foot-tall Heublein Tower in Simsbury's Talcott Mountain State Park was once the private home of food and liquor distributor Gilbert Heublein. He had promised his fiancée, Louise Gundlach, that he would build her a castle on a mountain. Completed in 1914, Heublein Tower was six stories high—more than enough to justify an elevator. And of course, it boasted of having the best elevator—a dependable Otis elevator.

Today, Otis is the world's top elevator and escalator manufacturing company. It estimates more than two million of its elevators are in active service, with about two billion rides taken each day. Currently, its headquarters is less than ten miles away from the Heublein Tower.

FLAT-BOTTOMED PAPER BAGS

Margaret E. Knight

In the 1870s, inventor Margaret E. Knight founded the Eastern Paper Bag Company in Hartford, Connecticut. It was based on her invention of a machine that allowed the inexpensive mass production of flat-bottomed paper bags, which eventually replaced most V-shaped bags. Her machine, which cut the paper, folded it and glued it, could perform the work of thirty people.

Knight's work as an inventor began early in life when, at age twelve, she witnessed an accident in a textile mill when a loom malfunctioned. That led to her invention of a widely adopted loom safety device. During her life, Knight was awarded patents on a machine to cut shoe soles, a "compound rotary engine," a window frame and sash and over twenty other inventions.

6 Sheets—Sheet 4.

M. E. KNIGHT.
Paper-Bag Machine.
No. 220,925.   Patented Oct. 28, 1879.

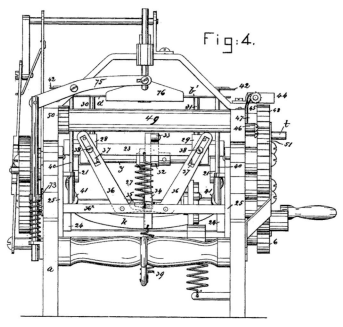

Fig: 4.

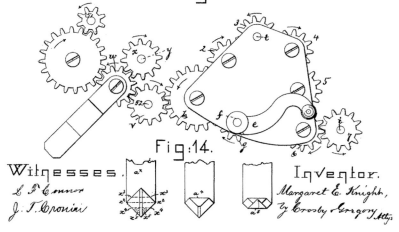

Fig: 13.

Fig: 14.

Witnesses.
L. F. Connor
J. T. Cronin

Inventor.
Margaret E. Knight,
By Crosby Gregory, Attys.

N. PETERS, PHOTO-LITHOGRAPHER, WASHINGTON, D. C.

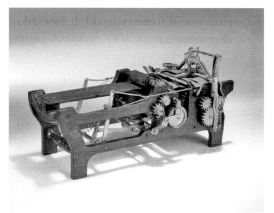

Opposite: An 1879 patent drawing of Margaret Knight's paper bag machine. *United States Patent and Trademark Office.*

Above, left: An 1879 patent model of Margaret Knight's paper bag machine. *United States Patent and Trademark Office.*

Above, right: One of today's flat-bottomed paper bags. In the late 1800s, this type of bag replaced most V-shaped bags. *Author's collection.*

When she passed away at age seventy-six, a local newspaper obituary referred to her as "a woman Edison." She was inducted into the National Inventors Hall of Fame in 2006.

POSTAGE METERS

Arthur H. Pitney and Walter Bowes

In 1902, Arthur Pitney was awarded a patent for the first postage meter. He formed a company to sell the device, but it wasn't until 1919, when he teamed up with England-born promoter and businessman Walter Bowes, that the machine took off. On March 15, 1920, the U.S. House of Representatives passed a law to allow "mechanical stamps" on first-class mail. A month later, Pitney and Bowes released their first product—the Model M Postage Meter, which printed two-cent postage. A few months later, Pitney Bowes's postage meter become the first one approved for use throughout the postal service. Within a few years, the postal services of Canada and the United Kingdom partnered with Pitney Bowes. The

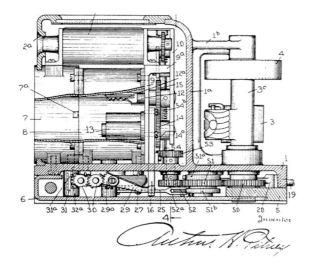

The drawing for one of Arthur Pitney's many patent applications. He and businessman Walter Bowes teamed up in 1919 to form one of the most successful companies in Connecticut history. *United States Patent and Trademark Office.*

Pitney Bowes Postage Meter Company's manufacturing operations were set up in Stamford, Connecticut.

The company introduced the first multidenomination postage meter in 1929. The device allowed postage to be applied to mail of various sizes and weights.

In the 1930s, the company invented postal equipment to handle packages of different weights and sizes. Even after Arthur Pitney's death in 1933, the company continued its innovations and aided the World War II efforts by inventing the API, a navigational instrument that provided continuous latitude and longitude readings. Walter Bowes retired as chairman of the board of Pitney Bowes in 1940 and died in 1957.

Today, Pitney Bowes is headquartered in Stamford, Connecticut, and its businesses include mailing, shipping, logistics, ecommerce and financial services.

FIRST PRACTICAL AUTOMATED FIRE SPRINKLER

Henry S. Parmelee

In 1874, Henry S. Parmelee of New Haven, Connecticut, who lost an arm while serving with the First Connecticut Volunteer Cavalry Regiment during the American Civil War, invented the first practical closed-head fire sprinkler. Each sprinkler head contained a bulb that, when damaged by the heat of a fire, would shatter, allowing the release of water. Parmelee installed the heads in his Mathushek Piano Company factory in New Haven. Later, many other companies adopted his sprinklers.

In addition to his six U.S. patents for fire extinguishing equipment, Parmelee was awarded patents for a railroad car brake (1883), a piano sounding board (1884) and an upright piano case (1885). He owned a piano factory and was also president of a streetcar company, the Fair Haven and Westville Street Railway Company. Today, the American Fire Sprinkler Association (AFSA) gives out its annual Henry S. Parmelee Award, which recognizes "individuals who have given extraordinary support to furthering the advancement and awareness of fire sprinklers."

Chapter 6

INDUSTRY

Although a small state, Connecticut is favored with rivers and streams that are perfect for water-powered manufacturing. It also has excellent harbors, such as those found in New Haven, Bridgeport and New London, that have made oceangoing transportation possible. The addition of highly skilled local people and talented immigrants from throughout the world has contributed to an environment where the Industrial Revolution of the nineteenth century flourished. The inventions of Connecticut people contributed greatly to industry's coming of age. In addition to the well-known inventions of Eli Whitney's cotton gin and Samuel Colt's firearms, thousands of now unknown residents invented and patented new machinery, improved existing devices and developed new manufacturing methods and procedures.

COTTON GIN

Eli Whitney

Two years after graduating from Yale College in 1792, Eli Whitney patented his cotton gin invention. It was a simple, but highly effective, machine that cleaned seeds from cotton. Manually removing the seeds proved to be the main bottleneck in the preparation and manufacture of cotton-based goods.

Whitney's invention, which used wires to separate the seeds from the cotton fibers, revolutionized cotton production in the United States and made it the main crop of the southern states. While a great boom for the economy, it had the horrible side effect of ensuring that many more people would be enslaved to produce cotton for domestic and foreign factories.

While Whitney's invention of the cotton gin was legendary, he spent years attempting to defend his patents in court and ended up with almost no profit. It was his use of mass-produced interchangeable parts that most affected manufacturing. Only a few years after his cotton gin invention, he developed a method of manufacturing muskets with interchangeable parts. It speeded up the manufacturing process tremendously and made Whitney a wealthy man.

After running his company in Hamden for many years, Eli Whitney passed away in New Haven in 1825.

Cotton gin and interchangeable parts inventor Eli Whitney. He established and developed Whitneyville, a manufacturing village in Hamden, Connecticut. *Library of Congress.*

INTERCHANGEABLE PARTS

Simeon North

Simeon North was often called the first official pistol maker of the United States and was also one of the country's first manufacturers of interchangeable parts, which made it easier and cheaper for an assembly line operation. He invented tools and machines that made individual parts, which were later assembled into finished products.

Starting off as a Berlin, Connecticut farmer and later mill owner (making scythes), in 1799, Simeon North received his first contract from the United States secretary of war—to manufacture guns for the military. Later, he built a second factory on the Coginchaug River in Middletown.

In 1813, for the War of 1812, North made five hundred flintlock horse pistols for the United States government. His 8,500-square-foot factory made as many as ten thousand pistols a year.

When Revolutionary War hero the Marquis de Lafayette made his famous visit to the United States in 1824 at the invitation of President James Monroe and the U.S. Congress, he visited Middletown, Connecticut. There, he was invited to meet Simeon North and see what was one of the largest factories in the country and one of the best examples of America's industrial advancement—Simeon North's facility in the Staddle Hill section of town. As stated in an early biography of North, the factory was "appointed with machinery elsewhere unknown" and was "one of the favorite points to which visitors to Middletown were taken."

In addition to his standardized, interchangeable parts innovations, North was often credited with creating the country's first milling machine—a device that uses rotating cutters to remove material from an object, usually made of metal, which is secured to a table-like surface. North's career lasted from 1799 until 1852. Shortly after he died in 1852, his company went out of business.

VULCANIZATION OF RUBBER

Charles Goodyear

Born in New Haven, Connecticut, and raised in Naugatuck, Charles Goodyear was a partner in his father's hardware business. When Charles was twenty-nine, the business failed, and he sought a new career. He began studying the possibility of improving natural India rubber so it would not be so adhesive nor so susceptible to changes in temperature.

A Massachusetts man, Nathaniel Hayward, had discovered that if rubber was treated with sulfur, it was no longer adhesive. Goodyear purchased the rights to Hayward's process. In 1839, Goodyear accidentally dropped rubber that had been mixed with sulfur on a hot stove and discovered it strengthened the rubber. He was awarded his first patent on the process—called vulcanization—in 1844. In his patent application with the title "Improvement in India-Rubber Fabrics," Goodyear wrote:

An 1860s photograph of vulcanized rubber developer Charles Goodyear. *National Portrait Gallery, Smithsonian Institution.*

My principal improvement consists in the combining of sulphur and white lead with the india-rubber, and in the submitting of the compound thus formed to the action of heat at a regulated temperature, by which combination and exposure to heat it will be so far altered in its qualities as not to become softened by the action of the solar ray or of artificial heat at a temperature below that to which it was submitted in its preparation—say to a heat of 270° of Fahrenheit's scale—nor will it be injuriously affected by exposure to cold. It will also resist the action of the expressed oils, and that likewise of spirits of turpentine, or of the other essential oils at common temperatures, which oils are its usual solvents.

He provides some details:

I take twenty-five parts of India-rubber, five parts of sulphur, and seven parts of white lead. The India-rubber I usually dissolve in spirits of turpentine or other essential oil, and the white lead and sulphur also I grind in spirits of turpentine in the ordinary way of grinding paint. These three articles thus prepared may, when it is intended to form a sheet by itself, be evenly spread upon any smooth surface or upon glazed cloth, from which it may be readily separated; but I prefer to use for this purpose the cloth made according to the present specification, as the compound spread upon this article separates therefrom more cleanly than from any other.

Legal challenges to Goodyear's rights in the United States occupied much of the next eight years. When much appeared settled, he traveled to Europe to find a place to manufacture his product but instead was met with other people who infringed on his rights to the vulcanization process. While he was overseas, other businesses in the United States began infringing on his patent.

During Goodyear's life, his rubber process was applied to a variety of industrial uses. After his death, it was applied to far more—from automobile tires to rubber balls, from rubber gloves to pencil erasers, from hoses to conveyor belts. Unfortunately, Goodyear did not benefit from it; he died deeply in debt at age fifty-nine.

In 1898, the Goodyear Tire and Rubber Company was named in his honor. It might be said that few discoveries in American history have resulted in as many practical uses as Goodyear's discovery. Goodyear was inducted into the National Inventors Hall of Fame in 1976.

PROCESS FOR WEAVING STRAW

Mary Dixon Kies

In 1809, Connecticut resident Mary Dixon Kies became the first woman to receive a U.S. patent in her own name. It was for a process to weave silk or thread into straw.

According to the National Inventors Hall of Fame, "Very little is known about her life or her accomplishments beyond her patented weaving process, which was widely used for over a decade" and "was adopted by the New England hat-making industry."

What is known is that Kies was born in South Killingly, Connecticut, which is a town only about two miles from the Rhode Island border. She, like many other women, learned that the 1790 Patent Act allowed "any person or persons" to petition for protection of their original methods and designs. Although women did not possess the same economic rights as men, they were apparently to be treated equally under that law.

Kies's invention was especially important at the time because U.S. president Thomas Jefferson had recently signed a law prohibiting the importation of British goods. The latter country was interfering with U.S. trade with Europe because Britain and Napoleon's France were in the middle of a war. Since many goods—including hats—couldn't be purchased from Britain, domestic manufacturers needed to fill the demand.

U.S. president James Madison's wife, Dolley Madison, sent a personal letter to Mary Kies congratulating her on her invention.

Sadly, Mary Kies didn't become rich from her invention. Fashion tastes changed, and she died in poverty at age eighty-four. Her grave wasn't even marked until the Killingly Grange did so in the 1960s. She was inducted into the National Inventors Hall of Fame in 2006.

Today, samples of straw fabric woven by Kies are in the Wadsworth Atheneum in Hartford and the Danielson Public Library in Danielson, Connecticut.

CORN-SHELLING MACHINE

Lester E. Denison

In 1839, Lester E. Denison of Saybrook, Connecticut, was granted a patent for a machine that shelled corn. Census records list him as a "wood turner" when he was about fifty years old and a "manufacturer" at about age sixty.

The Committee on Science and the Arts of Pennsylvania's Franklin Institute, an association that promoted the "mechanic arts," recommended Denison's invention for an award, as it was deemed "both new and useful." "Its operation…was highly satisfactory; the cobs were thrown out whole and divested of every grain, and the power requisite to keep it in action was less than in many other machines heretofore in use."

Within a couple of years after receiving his patent, Denison was granted two additional patents—another for a corn sheller and one for a burglar alarm. A quarter century later (1865), he received a patent for a photographic printing frame. In 1866, Denison died in the village of Winthrop, which is in the northwestern part of Deep River, Connecticut.

AMERICAN WINDMILL

Daniel Halladay

In the mid-1800s, Ellington, Connecticut businessman John Burnham asked machinist Daniel Halladay if he had a way to pull water from the ground. The result of Halladay's efforts was the first commercially feasible self-regulating windmill. His invention could automatically turn based on the direction of the wind. Also, the design of its sails allowed it to maintain a "uniform" speed. In a test, Halladay's machine pulled water from twenty-eight feet underground and transported it upward more than one hundred feet to the top level of a barn.

Halladay established the Halladay Windmill Company of Ellington, later moving it to South Coventry, which is about fifteen miles to the southeast. In 1863, its operations moved to Batavia, Illinois, to attract a wider market. As years passed, more windmill manufacturers moved to Batavia, and it became known as "The Windmill City."

Born in Vermont in 1826, Halladay eventually retired to California and died there in 1916.

STONE-CRUSHING MACHINE

Eli Whitney Blake

On June 15, 1858, New Havener and Yale college graduate Eli Whitney Blake, a nephew of cotton gin inventor Eli Whitney, received a patent on his "Machine for Crushing Stone" invention. It would greatly facilitate the construction of high-quality roads throughout the United States up through the twenty-first century.

In his patent application, Blake specified that his steam-powered machine may be "made of any size, varying according to the size of the stones it is to be capable of receiving and the amount of work it is to accomplish; and its proportions, having reference to strength, may be varied according to the hardness of the material on which it is to operate."

The idea for the machine had occurred to Blake when he observed the most common method of preparing stone for a new road—laborers breaking stones with hammers. Blake founded the Blake Rock Crusher Company to manufacture the stone-crushing machine. Within a couple of decades, five hundred of the machines were being used.

The two witnesses of Eli Whitney Blake's patent application were two of his brothers, Philos Blake and John Blake. Both were also his partners in a nearby door hardware factory. Less than two years after Eli Whitney Blake's stone-crushing machine invention, brother Philos would be awarded a patent of his own for a much smaller device—a corkscrew.

STEELMAKING

Alexander L. Holley

Alexander L. Holley was born in Lakeville, Connecticut (part of the town of Salisbury) in 1832. In 1863, Holley bought the American rights to British inventor Henry Bessemer's steelmaking process. Two years later in Troy, New York, Holley opened the first steelmaking plant in the United States using Bessemer's process and developing improvements to the process. In addition, he built the steelworks and rolling mills at Harrisburg and Pittsburgh, Pennsylvania, and at Chicago and Joliet, Illinois.

Of Holley's ten patented improvements to Bessemer's process, perhaps his most important was the invention of the "Holley Bottom." This was a removeable section underneath Bessemer converters that allowed the converter to remain hot when its brick bottoms needed to be replaced. A great deal of time and fuel costs were saved when the converters did not need to be restarted.

In Holley's later years, he headed up national civil engineering societies, wrote articles on phosphoric steel and the open-hearth steel process and lectured at the Columbia School of Mines.

ELECTRICAL PLUG AND THE PULL-CHAIN LIGHT SOCKET

Harvey Hubbell II

Harvey Hubbell II was born in New York in 1857 and moved to Bridgeport, Connecticut, when he was about thirty years old. He is credited with inventing the electrical plug and the pull-chain light socket.

The United States Patent Office's annual report for 1897 pointed out the incredible increase in the number of patented electrical inventions in the decades of the 1880s and 1890s. It noted that in 1880, 76 establishments with a total of 1,271 employees manufactured electrical apparatus and supplies. Those numbers increased by 1890 to 189 establishments with 9,485 employees.

On August 11, 1896, Harvey Hubbell of Bridgeport was awarded a U.S. patent for a pull-chain electric light socket, and on November 8, 1904, he received a patent for a "separable attachment plug." This latter device was designed so "that electrical power in buildings may be utilized by persons having no electrical knowledge or skill in the use of tools in attaching lights, fans, motors, heating apparatus, Surgical instruments, or any of the various appliances requiring the use of an electric current to fixtures in the circuit."

STEAM ENGINE INDICATOR

C.B. Richards

During the 1790s, James Watt invented the steam engine indicator to visualize pressure changes in low-speed steam engines. However, over the subsequent decades, many attempts were made to create an indicator that did a satisfactory job of measuring the efficiency of higher-speed engines. During the U.S. Civil War, thirty-year-old Charles Brinckerhoff Richards of Hartford, Connecticut, received a patent for an improved steam engine indicator that was able to more accurately record the efficiency of high-speed steam engines. Because of its effectiveness, its use spread throughout the United States and Britain.

Born in New York, Richards worked for Colt in Hartford for a few years during his early twenties, and he returned to Colt in 1861 to become its superintendent of engineering. Richards was instrumental in the design of Colt's Peacemaker revolver. Retiring from Colt, Richards helped found the American Society of Mechanical Engineers and became a professor and chair of mechanical engineering at Yale University for twenty-five years. He died in New Haven in 1919.

MANUFACTURING AUTOMATION SYSTEMS

H. Joseph Gerber

One of the most prolific holders of patents in Connecticut history was Jewish Holocaust survivor H. Joseph Gerber. Gerber was awarded more than six hundred U.S. and foreign patents during his lifetime. They included a numerical calculator, a machine that cut car seat covers and many other products for the optical, automotive, textile and other industries.

At the beginning of World War II, fifteen-year-old Gerber was a prisoner in a Nazi labor camp in Austria. The following year, he and his mother fled to the United States, settling in Hartford, Connecticut, in the 1940s. Within the space of a few years, he learned English, worked full time to support his mother and graduated from Hartford's Weaver High School. In 1945, he became an American citizen and the following year graduated from Rensselaer Polytechnic Institute with a bachelor's degree in aeronautical engineering.

In 1948, he founded Gerber Scientific, Inc. Its first product was a mechanical computational device named the Gerber Variable Scale. Immediately, sales to engineers and scientists took off. According to the National Museum of American History, it "consisted of two springs that expanded and contracted together to give proportional scales. These scales were used to multiply curves by constants and perform computations on graphs and curves to help reduce oscillograph and telemetry data."

From an investment of $3,000 in the late 1940s, Gerber personally headed up his company, which had $350 million in annual sales at the time of his death in 1996 in Hartford. Gerber introduced revolutionary products to a variety of industries. They included the GERBERcutter® System 70 (released in 1967), which was instrumental in the automation of the textile industry; the Gerber AM-5, an apparel CAD system (1980s); and AccuMark®, the first PC-based pattern making, grading and marker system. In apparel manufacturing, it was common to refer to Gerber as the industry's Thomas Edison.

In 1994, U.S. president Bill Clinton bestowed on Joseph Gerber the National Medal of Technology and Innovation (NMTI) "for his past and continuing technical leadership in the invention, development and commercialization of manufacturing automation systems for a wide variety of industries—most notably apparel—which have made those industries more efficient and cost effective in today's worldwide competitive environment."

Formerly located in South Windsor, Connecticut, Gerber Scientific's headquarters are now in Tolland.

Chapter 7

SCIENCE AND TECHNOLOGY

From Yale astronomer Ellen Dorrit Hoffleit, who catalogued hitherto undocumented stars, to Danbury's scientists and engineers who developed essential components for the Hubble Space Telescope, Connecticut has been at the forefront of astronomy. In the realm of computer technology, many of the best and brightest inventors and innovators had their start in the state.

A Litchfield, Connecticut native invented a half-tone photo-engraving process that allowed photographs to be printed on paper as easily and efficiently as text. Two Connecticut ophthalmologists developed the first practical method of iris identification, and a graduate of New Haven's Hopkins Grammar School, who received the first PhD in engineering in the country, was called "the greatest mind in American history" by Albert Einstein.

A man who grew up in Stratford was instrumental in establishing the first minicomputers; an engineer who was born in New Haven is credited with designing the first modern personal computer; and another New Havener has been called the "Father of the Internet" for his role in designing and implementing the Internet's basic communications protocols.

Today, new inventions are continuing at a rapid pace. For example, a Connecticut man's drone accessary invention has been responsible for saving the lives of swimmers who would have drowned if not for the device.

OLDEST CONTINUOUSLY PUBLISHED SCIENTIFIC JOURNAL

Benjamin Silliman

Yale College professor Benjamin Silliman founded what would become the oldest continuously published scientific journal in the United States. Begun in 1818, *The American Journal of Science and Arts* (later renamed *The American Journal of Science*) was often called "Silliman's Journal." Today, it is limited to geology and related sciences and is one of the most respected peer-reviewed journals in the world.

During his life, geologist and chemist Silliman was known for his forty-five-year-long editorship of the *Journal* and his distinguished tenure as a Yale professor. The silicate mineral sillimanite was named after him.

Silliman was born in the midst of the American Revolution in North Stratford, Connecticut, a month after the invasion of British troops in the 1779 attack on East Haven. He died during another American conflict—the Civil War—in North Stratford, sixteen days after Abraham Lincoln was reelected president of the United States. The first issue of *The American Journal of Science* to be released after Silliman's death stated, "His life...was passed in his native State, in connection with Yale College, the institution that early selected him as one of its faculty. Two or three times he was invited to become the president of colleges elsewhere, but New Haven continued his chosen home."

THERMODYNAMICS

Josiah Willard Gibbs

Born in New Haven, Josiah Willard Gibbs was the son of a linguist and theologian in Yale Divinity School. The younger Gibbs attended New Haven's Hopkins Grammar School (in later years, he would be a trustee and its treasurer) and was admitted to Yale College. At age nineteen, he graduated from Yale, where he won honors in mathematics and Latin. He continued on to receive the first PhD awarded in engineering in the United States. In 1866, Gibbs received a U.S. patent for an improved railway brake. In 1871, he was appointed chair of Yale's mathematical physics department, and he held that position for the next thirty-two years.

Gibbs's contributions to mathematics, physics and chemistry are based on his groundbreaking papers, which include the subjects of thermodynamical problems, the electromagnetic theory of light and other topics. His memoirs, titled *On the Equilibrium of Heterogeneous Substances*, helped to establish physical chemistry as a science. Gibbs was one of the earliest theoretical scientists in the United States to achieve international recognition, and Albert Einstein called him "the greatest mind in American history."

In 1901, Gibbs received the Copley Medal from the British Royal Society. Other than for infrequent trips, Gibbs lived his entire life in New Haven, and when he passed in 1903, he was buried in the city's Grove Street Cemetery.

In 1958, the United States Navy converted one of its ships into a 2,800-ton oceanographic research ship and renamed it the USNS *Josiah Willard Gibbs*. In 1964, scientists named a thirteen-kilometer-in-diameter moon crater the Gibbs Crater; later, the 2937 Gibbs, an asteroid discovered in 1980, was named after the famous scientist.

HALF-TONE PHOTO-ENGRAVING PROCESS

Frederic Eugene Ives

Born in Litchfield, Connecticut, in 1856, Frederic Eugene Ives worked as an apprentice at a local newspaper starting at age thirteen. He followed this up with experience apprenticing at printers in Ithaca and Greene, New York. At age eighteen, he applied for a position at the Cornell University photographic laboratory and was accepted. Ives later remembered, "I was so much interested in this experimental work that I slept in the laboratory, and worked at all hours, living principally upon crackers and milk. Once, I worked for a period of five days without sleep."

Ives was at Cornell less than four years, but they were incredibly productive. It was there he worked at solving a critical limitation of the traditional printing presses—that they were not capable of printing gray, they could only produce images in black-and-white. Images like photographs could only be printed in books, flyers and so on by using hand-engraved metal plates or wood blocks.

His invention solved this problem. Ives developed a "halftone process using a gelatin relief." His screen would convert an image into a pattern of dots. Large dots would be placed in dark areas and small dots in light areas. The result was the appearance of shades of gray. When an original photograph was shot through the Ives screen, it created a "half-tone" that could be engraved onto a metal plate, which in turn could copy it in ink onto paper just like it was text. This is much like the current process used to create images in newspapers, copy machines and laser printers.

In 1878, Ives left Cornell and joined a Philadelphia wood-engraving firm at which he could produce and sell his half-tone screens as well as other products. They manufactured his own invention—half-tone printing plates. As usual, he was continually working to improve his inventions. During his life, Ives was awarded about seventy patents.

Next on Ives's agenda was the creation of color photography. In 1885, he demonstrated a system of natural color photography at the Novelties Exposition of the Franklin Institute in Philadelphia.

In his 1894 *Hand-Book to the Photochromoscope*, Ives relates:

> *The attention of the author was directed to the subject as long back as 1878, but although much work was done, within the following years, very little real progress could be reported until 1888, when success was reached with the method as to quality of results in all of its applications. This was accomplished by perfecting the theory and practice of the negative-making process, so as to obtain the first true color records, by optical synthesis with colors representing the three fundamental color sensations, and by a system of clear gelatine printing and subsequent dyeing by immersion, for obtaining prints for synthesis by super-position....These improvements were followed by the invention of the Photochromoscope camera, in which the color-record is produced by a single exposure, on a single sensitive plate, and the Photochromoscope itself by which the synthesis is obtained without the trouble of making either color prints or screen projections.*

Ives used his Photochromoscope system to record his extensive travels throughout the United States and Europe.

In 1893, the Franklin Institute of Philadelphia, one of the most respected organizations dedicated to American science and technology research, concluded:

> *The Committee having carefully gone over the claims of Mr. Ives and his predecessors, and so far as able examined into their results, can come but to this conclusion: That Mr. Ives, by his original investigations and special construction of Camera and Photochromoscope for recording and*

reproducing colour, as set forth in United States Patent Specifications No. 432,530, July 22d, 1890; and No. 475,084, May 17th, 1892; has offered a practical solution of the problem of reproducing by means of photography the colours of nature; and therefore award to Mr. Frederic E. Ives the Elliot Cresson Gold Medal in recognition thereof.

During his life, Ives created many inventions connected with optics, printing and photography, including a short-tube, single-objective binocular microscope, the parallax stereogram—the first "glasses-free" 3-D display technology, the Photochromoscope camera and the chromogram, which combined and projected a three-separation color negative. This was a step toward full-color projectors.

In 1996, the U.S. Postal Service issued thirty-two-cent "Pioneers of Communication" commemorative stamps to honor Frederic Eugene Ives and three other late nineteenth-century pioneers in modern communications. Ives was inducted into the National Inventors Hall of Fame in 2011.

CATALOGING STARS

E. Dorrit Hoffleit

Working at Yale University from 1956 through her death in 2007, Ellen Dorrit Hoffleit was one hundred years old at the time of her passing.

As a senior research astronomer at Yale, she wrote the *Bright Star Catalog*. It included what are termed the "bright stars," which are the approximately 11,700 stars that can be seen without a telescope. Hoffleit included facts about each of these stars: their position, brightness, color, motion, velocity, history and so on. She specialized in the study of variable stars—those whose brightness changes. She classified these stars.

Born in 1907 to poor German immigrants on an Alabama farm, Ellen Dorrit Hoffleit's brilliance enabled her to be accepted for admittance to Radcliffe College, where she graduated in 1928. She applied for and was accepted as a research assistant at Harvard College Observatory. That led to her completing a master's degree in 1932 at Radcliffe with a thesis on the light curves of meteors. In 1938, she received her doctorate from Radcliffe College with a thesis on spectroscopic parallaxes, which means "determining the luminosities of stars, hence their distances, from line width and ratio diagnostics in their spectra." During World War II, she worked at determining the trajectories of antiaircraft missiles at the Ballistic Research Laboratory of the U.S. Army's Aberdeen Proving Ground.

Hoffleit received tenure as an astronomer at Harvard in 1948. In 1956, she accepted a research astronomer position at Yale University. She stayed there officially until her retirement in 1975 and unofficially for the rest of her life. While at Yale, she was concurrently director of the Maria Mitchell Observatory on Nantucket Island (from 1957 to 1978). It was there that she started a summer research program for undergraduate students.

While at Yale, Hoffleit wrote hundreds of scholarly papers, authored the book *Astronomy at Yale, 1701–1968* and wrote about eighty short news notes per year for *Sky and Telescope* magazine.

For Dr. Hoffleit's eightieth birthday, the Smithsonian Astrophysical Observatory's Minor Planet Center (MPC) named an asteroid after her. 3416 Dorrit is located about twice the distance from the sun as Earth, in the asteroid belt between Mars and Jupiter.

Apparently, Hoffleit's colleagues appreciated her human qualities as well as her intellectual professionalism. A comprehensive 2007 obituary of her by Dr. Virginia L. Trimble, professor of physics and astronomy at the University of California–Irvine, ends with, "Part of what made it so wonderful to encounter her was not just that you were glad to see her, but that she was glad to see you."

THE HUBBLE SPACE TELESCOPE

The 43.5-foot-long Hubble Space Telescope was deployed from the space shuttle *Discovery* on April 25, 1990. Powered entirely by the sun, it orbits the Earth at an altitude of about 340 miles, traveling at about 17,000 miles per hour. Without the obstructions of light pollution, clouds and atmospheric distortions, it has been able to allow mankind to see far further into space than any earth-based telescope could ever see—to observe distant stars, galaxies and other astronomical objects. It has become the National Aeronautics and Space Administration's (NASA) longest-running project.

Many of the most important components of the Hubble were designed and built by optics designers and engineers at Perkin-Elmer Corporation's Danbury, Connecticut facility, which is perched on a hill overlooking Route 7. Some of these components include the mirrors and guidance sensors. Its primary mirror is 94.5 inches in diameter and weighs 1,825 pounds; the secondary mirror is 12.2 inches in diameter with a weight of 27.4 pounds.

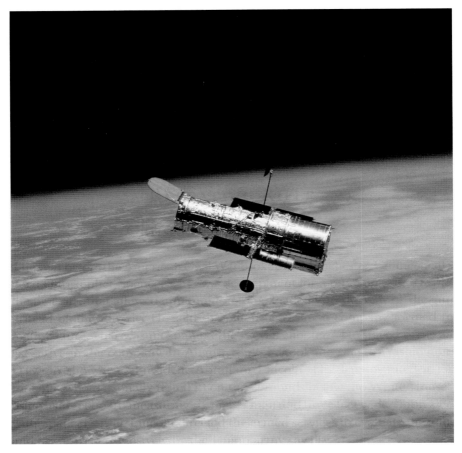

Key components of the Hubble Space Telescope were designed and built by Connecticut's Perkin-Elmer Corporation. *National Aeronautics and Space Administration.*

A tiny flaw in the primary mirror was discovered after deployment. Although the mirror was perfectly polished, it had an aberration that affected the clarity of its images. The problem was fixed by astronauts who installed corrective optics in December 1993. Since then, the telescope has been doing an outstanding job of fulfilling its mission. With a range of light that extends from the ultraviolet into the near-infrared, it has made more than 1.5 million observations. Each year, Hubble transmits back to Earth about ten terabytes of new data. About twenty thousand peer-reviewed science papers have resulted from its findings.

Since 1990, the Hubble Space Telescope has been making an orbit around the Earth every ninety-seven minutes. *National Aeronautics and Space Administration.*

Today, as it has done for over three decades, the space telescope makes a complete orbit around the Earth every ninety-seven minutes. Over its lifetime, five astronaut servicing missions (in 1993, 1997, 1999, 2002 and 2009) have resulted in extending the telescope's useful life from an initial estimate of fifteen years to thirty years and counting. In addition, they have delivered and installed some of the most advanced scientific instruments.

The Hubble Space Telescope has far exceeded most expectations. As explained on the NASA website, Hubble "has tracked interstellar objects as they soared through our solar system, watched a comet collide with Jupiter, and discovered moons around Pluto. [It] has peered back into our universe's distant past, to locations more than 13.4 billion light-years from Earth, capturing galaxies merging, probing the supermassive black holes that lurk in their depths, and helping us better understand the history of the expanding universe."

Many people see the Hubble Space Telescope as NASA's greatest achievement—with the sole exception of the Apollo missions to the moon. It is difficult to overstate the dedicated professionalism that Connecticut's people had in order to achieve a successful outcome and add to our knowledge of outer space.

PHOTOFLASH SYNCHRONIZER

Morris Schwartz

In 1930, Russian-born Morris Schwartz invented the first photoflash synchronizer. It was named the Kalart Flash Synchronizer after his father's name—KALman SchwARTz. It synchronized the firing of a flash with the opening of the shutter allowing light to photographic film. This allowed the capture of high-speed flash photos of moving objects in poor lighting, such as at indoor sporting events.

After working for the *New York Times* as a darkroom operator and as a photographer at the *Jewish Daily Forward*, which was known as being the most popular Yiddish-language newspaper in the United States, Schwartz began manufacturing his photoflash synchronizer, as well as other inventions, in New York City. In the early 1940s, he moved his operations to Stamford, Connecticut, and then to Plainville, Connecticut. Employing over three hundred people, he made Kalart Cameras, movie editing equipment, 16mm movie projectors and 8mm editing and movie projectors.

Schwartz died in 2004 at age 103. During his life, he was awarded over one hundred U.S. patents for photo and audiovisual inventions.

COLOR PHOTOGRAPHIC FILM

Leopold Godowsky Jr.

As teenagers, Leopold Godowsky Jr. and Leopold Mannes, the sons of famous musicians, experimented with ways to produce color photographs. While entering careers in music, they continued to collaborate on experiments in photography until 1930, when they began working for Eastman Kodak. Their work there led to the development of Kodachrome® color film in 1935 (motion pictures) and 1936 (still photographs). The two men used thin photography film that was coated with five layers of emulsion. Godowsky, long a Westport, Connecticut resident, continued his work on color film in his Westport laboratory for many years.

A classical concert violinist by profession, Godowsky performed for the San Francisco and Los Angeles Symphonies. (His father was world-famous pianist Leopold Godowsky, and his wife, Frances, was the sister of George and Ira Gershwin.)

A student at Juilliard and Harvard Mannes, Godowsky's co-inventor, Mannes (1899–1964), became a successful pianist and composer and was president of the Mannes College of Music. Both Godowsky and Mannes were inducted into the National Inventors Hall of Fame in 2005.

Due to the introduction of digital photography, Kodak ceased production of Kodachrome® film in 2009. However, the thousands of motion pictures produced using Kodak's process and the countless billions of still photographs that survive in personal collections mean that Kodachrome's place in history isn't forgotten.

Iris Identification
Leonard Flom and Aran Safir

In the 1970s, Fairfield ophthalmologist Leonard Flom had the idea that images of the human iris could be used in the identification of people in much the same way that fingerprints had been used since the nineteenth century. Throughout the 1980s, he and his colleague Aran Safir worked on the invention of an iris identification system. In 1964, Safir had already been awarded a patent (U.S. Patent No. 3,136,839) for an early electronic retinoscope, which is an instrument for observing a subject's retina to determine its state of refraction.

In 1987, Flom and Safir received a U.S. patent for an iris identification system (U.S. Patent No. 4,641,349). Based on the fact that every person's iris is unique (including those of identical twins), they proposed a way to identify individuals by the structure of their eyes. Their patent application reveals their methods and the apparatus that illuminates the eye "until the pupil reaches a predetermined size, at which an image of the iris and pupil is obtained. This image is then compared with stored image information for identification."

Today, iris recognition is considered by most experts to be the most accurate method of biometric identification. Its accuracy, ease of use and noncontact nature have ensured its adoption for a variety of purposes.

In later years, Dr. Safir was a member of the Mount Sinai School of Medicine faculty and director of ophthalmology at the University of Connecticut.

Dr. Flom co-founded the company IriScan, which made iris-scanning cameras. Years after the receipt of his patent, Flom developed a system of taking images of the eyes of newborn babies and their birth mothers while they are still in the delivery room. The images then are checked at the time of their discharge from the hospital.

In his later years, while living in Westport, Flom traveled to Assaf Harofeh Hospital in Ramla, Israel, for research and taught at the New York University School of Medicine, where many decades before he had received his medical degree. He also was a member of Congregation Beth-El in Fairfield and served on the Ethics Committee of the Connecticut Medical Examining Board.

Flom and Safir were inducted into the National Inventors Hall of Fame in 2013. Aran Safir died in 2007 at age eighty. Leonard Flom died in 2022 in Norwalk at age ninety-four.

CMOS ACTIVE PIXEL IMAGE SENSOR

Eric R. Fossum

Eric R. Fossum, currently a professor of engineering at the Thayer School of Engineering in New Hampshire, was born in Simsbury, Connecticut. He graduated from Trinity College in Hartford in 1979 with a bachelor's degree in engineering and obtained his master's and doctorate in engineering and applied science from Yale University in 1980 and 1984, respectively. After working in Columbia University's Electrical Engineering Department, he moved to NASA's Jet Propulsion Laboratory in 1990. It was there that he invented the "CMOS active pixel image sensor and the Camera-on-a-Chip CMOS image sensor."

Fossum has won the NASA Exceptional Achievement Medal (1996), the Royal Photographic Society Progress Medal (2004), the Yale University Wilbur Cross Medal and the Society of Motion Picture and Television Engineers Camera Origination and Imaging Medal (both 2014) and shared the Queen Elizabeth Prize for Engineering (2007) with Dr. George Smith (charge-couple device or CCD), Professor Nobukazu Teranishi (developed photon counting image sensors for visible light X-ray) and Dr Michael Tompsett (imaging semiconductor circuit and the analogue-to-digital converter chip) for their work on digital imaging sensors.

The Queen Elizabeth Prize for Engineering Foundation stated:

> *Together, this image sensor technology has transformed medical treatments, science, personal communication and entertainment—from Skyping, selfies,*

computer games and feature length digital movies—to reporting live from wars using the small camera on a smartphone. It saves lives, by using non-surgical pill cameras and endoscopes inside our bodies to diagnose medical problems, as well as helping to reduce X-ray doses to patients and improving dental care.

Image sensors inside cars increase driver safety, enhance security on the streets and expand our knowledge of the universe through images from the surface of Mars or a comet, from spacecraft in orbit around other planets and the breathtaking pictures of some of the billions of galaxies surrounding the Milky Way.

Minicomputers

Kenneth H. Olsen

Kenneth H. Olsen, the founder and former CEO of the Digital Equipment Corporation, was born in Bridgeport, Connecticut, in 1926 and grew up in nearby Stratford. One of his early hobbies was repairing radios. During World War II, he served as an electronic technician in the U.S. Navy. He then earned his bachelor's and master's degrees in electrical engineering from the Massachusetts Institute of Technology (MIT) in 1950 and 1952, respectively.

In a 1988 interview with the Smithsonian Institution's National Museum of American History, Ken Olsen spoke of his early years:

I was brought up in Connecticut, outside of Bridgeport. It was an area where machine tools were built, where you were normally expected to learn machine shop practice. And I did. But there wasn't much in the way of electronics going on. When I was drafted for World War II, I had the enormous opportunity to go to electronics school in the Navy. It was a great school. It lasted a year, or at least eleven months. It was set up by competent people and they gave an excellent education in electronics. They taught us all the tricks, manipulating, calculating circuits, the rules of thumb for electronics, and went through all of the gimmicks and tricks and things one should know about radio. And then radar, and counter measures. It was the most exciting thing a young kid…could go through.

After obtaining his master's degree, Olsen stayed on at MIT at its Lincoln Laboratory, worked "on the development of computers for the SAGE Air

Defense System" and "contributed to the development of Jay Forrester's Magnetic Core Memory, and built his first computer."

During the 1950s and 1960s, huge mainframe computers were standard in the business world. Olsen decided to develop a smaller, more affordable interactive computer. He especially envisioned the benefits these machines would offer scientists and researchers.

Olsen left Lincoln Laboratory in 1957 to form the Digital Equipment Corporation (DEC) with fellow engineer Harlan Anderson. Olsen would serve as its CEO for the next thirty-five years. DEC's first computer hit the marketplace in 1960. Named the PDP-1, it sold for $120,000 at a time most mainframes cost over $1 million. It was the first of what came to be called "minicomputers."

DEC's growth continued until by the late 1980s, with 120,000 employees, it was the second-largest computer company in the world. Only IBM was larger. In 1993, Kenneth Olsen received the IEEE Founders Medal from the Institute of Electrical and Electronics Engineers "for technical and management innovation, and leadership in the computer industry."

The late 1980s brought an unwelcome turn in DEC's fortunes with the advent of the microcomputer. Apple had introduced its first personal computers in the 1970s, and IBM broke into the business with its 1981 IBM PC personal computer. With Microsoft's software operating system available, many companies began producing what were called IBM clones.

In 1998, the Digital Equipment Corporation was acquired by COMPAQ Computer Corporation, which was a leading manufacturer of the PC clones. In an article for historyofyesterday.com, Barry Silverstein lists some of the positive ways that DEC advanced computer technology:

- *The company was responsible for the commercial implementation of Ethernet*
- *The first versions of the computer language C and the UNIX operating system ran on DEC's PDP series*
- *Many of DEC's operating systems, such as OpenVMS, were widely used*
- *A proprietary network called DECnet led to one of the first peer-to-peer networking standards*
- *DEC invented clustering, in which multiple machines operated as one*
- *DEC developed the innovative and powerful 64-bit Alpha series of microprocessors*
- *DEC developed Notes-11, one of the first examples of online collaboration software, or groupware*

- *AltaVista, the first truly comprehensive Internet search engine, was developed by DEC, well before Google was introduced*

Kenneth Olsen passed away in 2011 at the age of eighty-four. Upon his death, Microsoft Corporation founder Bill Gates said in a letter to Gordon College, "An inventor, scientist, and entrepreneur, Ken Olsen is one of the true pioneers of the computing industry. He was also a major influence in my life and his influence is still important at Microsoft through all the engineers who trained at Digital and have come here to make great software products."

FIRST MODERN PERSONAL COMPUTER

Wesley A. Clark

While most of the people in this book worked on their greatest inventions or innovations while living in Connecticut, some people's only claim to being a Connecticuter is their birth. One such person is Wesley A. Clark. Born in New Haven in 1927, he grew up in the states of New York and California. When he passed away at age eighty-eight, his *New York Times* obituary declared he "designed the first modern personal computer" at the Massachusetts Institute of Technology in the 1960s.

Long before the days of Apple and Microsoft, Clark advocated research on computers that could be owned and used by single users rather than large mainframes that would be shared by many users. In the 1950s at MIT's Lincoln Lab, he was one of the creators of the first fully transistorized computer—the TX-0. In 1961, he (along with his associate Charles Molnar) headed up a team that developed the LINC (Laboratory Instrument Computer) transistorized minicomputer, which was developed for medical and scientific applications. With a simple operating systems, tiny screen and magnetic tape storage, it had a one-half-megahertz processor and when commercially released in 1964 cost over $40,000. Most important of all, it was designed as a stand-alone system to be used by one individual—the very essence of a "personal computer."

Later, Clark said, "What excited us was not just the idea of a personal computer. It was the promise of a new departure from what everyone else seemed to think computers were all about."

INTERNET'S BASIC COMMUNICATIONS PROTOCOLS

Vinton Cerf

Vinton Cerf is frequently called the "Father of the Internet" because he was instrumental, along with electrical engineer Robert Kahn, in designing and implementing the Internet's basic communications protocols, TCP/IP (Transmission Control Protocol/Internet Protocol). This protocol allows any computer "with the appropriate connection to enter" the network.

Cerf was born six weeks prematurely in New Haven, Connecticut, in 1943. The procedure at that time was to place premature babies in oxygen tents to help them breathe. According to Cerf, it's believed that this might have caused his progressively deteriorating hearing impairment. He began his lifelong need for hearing aids in both ears at age thirteen.

In the 1960s, Cerf worked for the United States Department of Defense Advanced Research Projects Agency (DARPA) on the development of the Internet. Later, in working on the ARPANET project, he was personally attracted by the idea of the development of email because it could replace telephone communications, which are often a problem for people with hearing disabilities.

With his own experience in mind, Cerf has recently said, "It's a high priority to ensure that disabilities do not prevent people from gaining the full benefit of online and offline digital environments." He was adamant that there is no excuse for making products that are not accessible."

Among his many awards, Cerf has been inducted into the National Inventors Hall of Fame and has received the National Medal of Technology from President Bill Clinton; the Presidential Medal of Freedom from President George W. Bush; the Queen Elizabeth Prize for Engineering in 2013; and the French Legion of Honour.

LIFESAVING DRONES

Bill Piedra

Longtime Connecticut resident Bill Piedra developed a life preserver that can be dropped by a drone onto a person who is in danger of drowning. Designed to help keep people afloat until lifeguards can reach them, the drone accessory is titled Project Ryptide. It's an appropriate name since a riptide is a strong current flowing away from a shore that presents a hazard to swimmers and boaters. Each year, about one hundred people are killed by riptides in the United States alone.

An article from news station WTNH quoted Piedra as saying, "A volunteer or secondary member can get this device out to a person in water 500 yards away in less than 30 seconds." After the life preserver touches the water, it automatically inflates.

Bill Piedra moved to Connecticut when he was about ten years old and has lived in Stamford and Manchester. He had a long career as an inventor and software developer and explained in a recent interview for this book that he had the idea of a lifesaving drone since he was a little kid. It would turn into the hobby of radio-controlled airplanes.

During his time as a private contractor, Piedra had as many as seventy-two full-time employees working for him. In 2014, he sold his company and was "trying to retire." He says, "The idea stuck in my head from years before: Let's attach a life preserver to a drone and try to drop it in water." At the time, he was mentoring Stamford high school students and asked if they would like to make his invention a reality. Four students enthusiastically agreed and set to work creating the Piedra lifesaving drone. It consisted of a drone with a life preserver attached to its bottom. When a button on a remote control is pushed, the life preserver with an attached CO_2 canister is dropped into the water close to the person in trouble. When it touches the water, a green tablet on it dissolves and fires a spring. That spring pokes a hole in the CO_2 cartridge and inflates the life raft in about five seconds.

Piedra and the students tested the lifesaving drone; its performance was impressive, and soon it was picked up by print and broadcast media. The inventor and his students were invited on almost every national morning TV show in the United States. Then, news of the device spread throughout the world.

In August 2016, the summer Olympics were held in Rio de Janeiro, Brazil. Bill Piedra was asked by General Electric Corporation to fly down to Brazil and give demonstrations of his drone. Today, he remembers Brazil as having the "most beautiful beaches on the planet Earth." He met members of the Rio lifesaving group called Sobrasa. Every couple of thousand feet along the beaches, they had a lifesaving station. Whenever someone was in trouble, thirteen or fourteen lifeguards would rush out into the ocean to try to save them. Piedra's invention was a perfect addition to their array of rescue equipment.

In 2016, on the Science Channel TV show *Droned*, Piedra was originally hired as a technical consultant for the first episode. They liked his appearance so well that they made him a cast member, and he appeared in all ten episodes.

Today, Bill Piedra's lifesaving drone technology is known and used throughout the world. In addition to water rescues, it has played a vital part in ice rescues and boat rescues. Sometimes when he is traveling, Piedra will mention that he is the inventor. Not surprisingly, people sometimes respond, "Yeah, yeah, sure you are." Piedra is not sure how many lives have been saved with this lifesaving device dropped from a drone, but it sure makes sense for beaches to have them!

Chapter 8

SPORTS AND LEISURE

J ust like in every other area of life, Connecticut people made their mark on the world of recreation. One New Britain native was more responsible than anyone else for the invention of American football; a Hartford man designed the chest protector that every professional baseball catcher wears; and a man living in Fairfield invented a plastic baseball that has been used by just about every American youngster during the past six decades.

In the area of gaming, a New Yorker invented Scrabble, but it took a Connecticut couple to give it a name and to manufacture and sell it. A Bridgeport-born physicist invented the first interactive analog computer game. Even the writer and humorist Mark Twain got into the act and invented and patented a board game that would educate players on world history.

MARK TWAIN'S BOARD GAME

In the 1880s, while living in Hartford, one of the most famous writers in America invented a history board game. In his 1884 patent application, filed under his real name, Samuel L. Clemens, he described the object of his invention: "to provide a new and improved instructive and amusing game." As the game progressed, players were expected to learn key dates and facts from world history. When it was first sold in 1892, it was titled Mark Twain's Memory Builder.

An 1890 painting of writer Mark Twain (Samuel L. Clemens) by Charles Noel Flagg. *The Metropolitan Museum of Art, New York.*

In the game, points were earned for correct answers to questions on subjects like the dates of U.S. presidencies, battles and inventions. Twain explains the game: "The first player takes a pin and states that he will place it in the hole of a certain year, and at the same time he states an event of that year. If his statement is correct, he places the pin in the corresponding hole, and if his statement is not correct he is not permitted to place the pin in the hole." Twain recommends an umpire who would decide if the statements are correct.

Discussing ways the game can be educational for adults, Twain stated:

> *Often one knows a lot of odds and ends of facts belonging in a certain period but happening in widely separated regions; and as they have no connection with each other, he is apt to fail to notice that they are contemporaneous; but he will notice it when he comes to group them on his game-board. For instance, it will surprise him to notice how many of his historical acquaintances were walking about the earth, widely scattered, while Shakespeare lived. Grouping them will give them a new interest for him.*

Regarding child players, Twain wrote, "Many public-school children seem to know only two dates—1492 and 4ᵗʰ of July; and as a rule they don't know what happened on either occasion. It is because they have not had a chance to play this game."

THE WIFFLE® BALL

David N. Mullany

Born in Massachusetts in 1908, David N. Mullany grew up on a tobacco farm near the Connecticut River. His favorite sport was baseball, which he played in high school and during his years as a student at the University of Connecticut. Graduating at the beginning of the Great Depression, he sought work in Bridgeport, Connecticut. While job hunting, he pitched for a local baseball league. When he was finally offered a job, it was for a local patent drug manufacturer that not only wanted a good employee but also needed a good pitcher for its company baseball team.

About two decades later, David Mullany, then living in Fairfield, Connecticut, noticed that his son David and his friends had a problem. They had been using tennis balls as baseballs and damaging neighborhood windows and siding. After numerous complaints, they switched to practice golf balls, which were made of plastic. Everything worked out OK until the boys developed sore arms trying to throw curveballs with the small plastic balls.

Mullany had an extensive career in business and was familiar with products that were packaged in spherical plastic "eggs." He figured that if

University of Connecticut graduate David N. Mullany invented the WIFFLE® ball in the 1950s. *The Wiffle Ball, Inc.*

such a package was shaped like a baseball, it could be used as a practice ball. He obtained some samples of hollow plastic balls and went to work cutting them in different ways—trying out each design in his backyard. One design worked: the plastic ball that was composed of one-half with eight oblong holes and the other half without any holes.

Mullany was sure the ball that he invented could prevent the great majority of damage done by baseballs and softballs during children's play. As David J. and Stephen A. Mullany wrote in their *WIFFLE® History*, their grandfather was so confident consumers would love the ball that he

> *took out a second mortgage on his house, borrowed from friends and went ahead with forming a new toy company. But what to call the product? "Mullany ball" didn't have much of a ring to it and Gramp wasn't an egomaniac. He was pretty low key and shied away from the spotlight and publicity. He talked some more with dad and asked what they called the game. Dad said one word "Wiffle." The game played was a pitcher's duel with lots of strikeouts. In dad's neighborhood, when you took a swing and missed, you whiffed! So it was decided, they'd call the product a "WIFFLE" ball. (No "h" in "Wiffle" would mean one less letter they'd have to pay for on the sign.) The name stuck and it's our brand name and the federally registered trademark that we use today.*

Today, there are many copycats—but only one authentic Wiffle® ball.

Many of baseball's greatest major-league players look back fondly on how they grew up with Wiffle® balls. An article on the Society for American Baseball Research's website mentions New York Yankee great Don Mattingly growing up with backyard Wiffle balls.

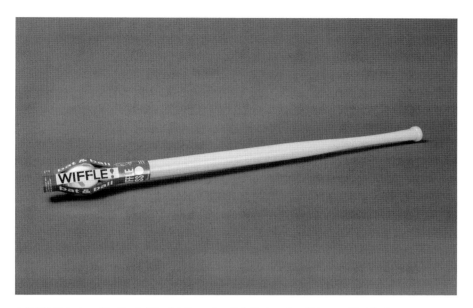

A Wiffle® bat and ball set. The Mullany family's factory is still located in Shelton, Connecticut. *The Wiffle Ball, Inc.*

Years later, he recalled those games, saying, "I can't imagine what my parents must have been thinking, with that [Wiffle] *ball banging against the metal door every two minutes."* [Wiffle] *ball helped Don develop his ability to drive the ball the other way. On the backyard field a thickly-leaved tree hung over the field on the first-base side. If you hit the tree, you were out. In left field, however, Mattingly recalled, there was the family garage. A fly ball onto the garage was counted as a home run. Mattingly learned pretty quickly to hit the ball the other way. It was a skill that served him well for many years to come.*

After manufacturing their successful ball, it became apparent to the Mullanys that selling a bat along with the ball made financial sense, so WIFFLE® bats were added to their product line. Eventually, inventor David N. Mullany was replaced by his son David A. as head of the company. He, in turn, was followed by his sons, David J. and Stephen A. Mullany. Since their inception, Wiffle Balls® have been manufactured in only one place—at the Mullany family's factory located in Shelton, Connecticut, which is still owned and operated by the family after more than sixty years.

FRISBEES

Sports involving the throwing of discs have been around for thousands of years. Most notable were the discus throw competitions in the Olympic Games of ancient Greece. That disc was made of stone, iron, lead or bronze and weighed up to about twelve pounds. In the twentieth century, however, the most famous disc throwing was by ordinary people, and the discs were made of plastic or thin metal.

Many authorities say that frisbees began when Yale University students first threw empty pie plates from Mrs. Frisbie Pies across the New Haven Green. *Author's collection.*

The Connecticut government's official state website, CT.gov, has a web page on "Connecticut's Historical Facts." Under the year 1920, it states, "First Frisbee, Yale students discovered empty pie plates from Mrs. Frisbie Pies in Bridgeport could be sailed across the New Haven Green." This led to students shouting the word "Frisbee" to warn other people that they had just launched a disc in their direction. The original plates were made of tin and embossed with the words "Frisbie's Pies" in large print. The company that made Mrs. Frisbie Pies was founded in 1871 and discontinued operations in 1958.

In 1948, a plastic throwing disk called the Flying Saucer was invented, and later it was modified and sold to the Wham-O company as the Pluto Platter. Both names were obviously inspired by the flying saucer sightings of the 1950s. It has been estimated that over 100 million of the modern plastic throwing discs were sold before 1977. Wham-O was bought by the company Mattel Toy in 1994.

Today, scores of companies throughout the world make frisbee-like discs, and frisbee games and competitions are held all over the United States—and the world.

SILLY PUTTY®

James Wright

During World War II, the Japanese invasion of Asia cut off much of the United States' supply of rubber. In response, chemist James Wright attempted to develop a synthetic rubber at a General Electric laboratory in New Haven in 1943. He discovered that combining boric acid with silicone oil created a stretchy, bouncy material. It wasn't a rubber substitute, and it apparently wasn't good for much.

All that changed in 1950, when advertising expert Peter Hodgson took some of it to the International Toy Fair in New York, where it was a success. To improve its marketability, Hodgson named it Silly Putty® and placed it in plastic egg-like containers. In addition to its use as a toy—it was capable of being bounced and of picking up comic book images and text—it served as way to pick up lint, pet hairs and other small particles. It also has been used in science and medicine.

In 1968, Silly Putty went on the Apollo 8 mission to the moon. Astronauts Borman, Lovell and Anders used it to fasten down their tools as they became the first human beings to orbit the moon. Today, the name Silly Putty® is a trademark of Crayola LLC.

ERECTOR SET

Alfred Carlton "A.C." Gilbert

For much of the twentieth century, the A.C. Gilbert Company was New Haven's largest employer. From 1909 until the company went out of business in 1967, its products of children's toys and educational kits included chemistry sets, electrical sets and optics and physics sets; Gilbert had more than 150 patents during his career.

However, the company's biggest claim to fame was the Erector Set, which was invented by its founder and former gold medal–winning Olympic pole-vaulting champion, Alfred Carlton Gilbert. The product began in 1911, when the twenty-seven-year-old Gilbert observed workers riveting steel beams. Two years later, Gilbert received a U.S. patent for his "Toy Construction Blocks." When the product was first sold, it was under the name Mysto Erector Structural Steel Builder.

A.C. Gilbert introduced Erector® at New York City's Toy Fair in 1913. All parts were stored in unpainted wooden boxes and had 1¼-inch girders that children used for construction. By 1935, Gilbert had sold more than thirty million sets, including a motorized merry-go-round, truck, Ferris wheel, Zeppelin, steam shovel and others. A.C. Gilbert Erector Sets sold until 1962, when Meccano Company (now Meccano Toys Ltd.), a British toy company, acquired it. It now markets construction and racing vehicles. One kit, the Meccano Super Construction 25-in-1 Motorized Building Set, includes over six hundred parts and can be used to build a crane, a car, a helicopter and more. Vintage A.C. Gilbert Co. Erector Sets can be quite valuable—a mint condition 1950s set may be worth $2,000.

SCRABBLE

James Brunot

New York–born architect Alfred Mosher Butts invented the board game Scrabble in 1938. However, without experience in business, the game, which he named Criss-Cross Words, went nowhere—his patent application was turned down, and no company was interested in manufacturing it.

Butts, while retaining patent rights, then sold the right to produce the game to a Connecticut friend—social worker and part-time sheep farmer James Brunot. Brunot disliked his two-hour-long daily commute from Connecticut to his job in New York City and wanted to start a Connecticut business. He proceeded to change some relatively minor features of the game, renamed it Scrabble and aggressively marketed it.

Brunot and his wife, Helen, began manufacturing Scrabble games in their home, but demand was so overwhelming that they outgrew it; later, Brunot explained there was "no room for anything but boxes and racks and tiles." In 1948, Brunot received a copyright and patent for the game, and he and Helen set up a factory in a former schoolhouse in Dodgingtown, which is part of Newtown. Connecticut. By 1952, sales were still mediocre—about ten thousand sets a year—but due to word of mouth and important orders, especially a huge one from Macy's, close to one million sets were sold the following year.

As the Brunots tried to satisfy orders for hundreds of thousands of Scrabble games a year, they realized that they needed to sell the business—or at least the bulk of it. They proceeded to sell the rights to the game in the United States and Canada to a major game manufacturer, except for a deluxe version of the game that they retained to keep the company alive.

Helen and James Brunot died in Bridgeport, Connecticut, in 1972 and 1984, respectively. Scrabble's inventor, Alfred Mosher Butts, passed away in 1993 at age ninety-three.

Today, with over 150 million games sold worldwide, Scrabble is the most popular word game in history.

Video Games

William Higinbotham

William Alfred Higinbotham, who became one of the most famous physicists of the twentieth century, was born in Bridgeport, Connecticut, in 1910, the son of a Presbyterian minister. During World War II, he was one of the Manhattan project scientists who developed electronic components for the first nuclear bomb at Los Alamos National Laboratory in New Mexico. After the war, Higinbotham worked hard to stop the spread of nuclear weapons. He was a founder of the nuclear nonproliferation organization Federation of American Scientists and served as its first chairman.

After the Brookhaven National Laboratory was founded in New York in 1947 to conduct research in the atomic sciences, Higinbotham joined because of his unequaled expertise and because the organization would not interfere with his activism in the cause of nuclear arms control. In addition to researching peacetime nuclear power, Brookhaven was involved in developing computer technology.

In 1958, Higinbotham created a milestone in computer gaming with his Tennis for Two, which was the first interactive analog computer game. Using an analog computer, handheld controllers and an oscilloscope with a five-inch-diameter screen, it was also the first computer game to show motion. Many years later, Higinbotham remembered how he created the game: "It took me about two hours to rough out the design and a couple of weeks to get it debugged and working."

Tennis for Two's place in computer history has led to Higinbotham being called the "Grandfather of Modern Video Games." Although he received more than twenty patents in his career, he never patented Tennis for Two; since he created it while working on government projects, he figured the government would own it. Besides, for the last half century of his life, he most wanted to be known as someone who fought to stop the spread of nuclear weapons and prevent a nuclear war. Higinbotham passed away in 1994 at age eighty-four.

AMERICAN FOOTBALL

Walter Chauncey Camp

The person most responsible for inventing American football was Walter Camp, a native of New Britain, Connecticut. An undergraduate and medical student at New Haven's Yale University from 1876 to 1881, he was a halfback and captain of its rugby-like football team. Camp was instrumental in setting the rules of modern football while serving on the Intercollegiate Football Association's Rules Committee for forty-eight years.

Following his recommendations, the organization set team size at eleven men, established the position of quarterback and introduced the requirement that a team would surrender possession of the ball after not achieving a specific number of yards in what they called "downs." Also, they established a line of scrimmage, offensive signal calling and the scoring system. During the years 1888 to 1892, Camp coached the Yale football team to sixty-seven wins with only two losses.

Camp was the first person to publish a book on American football and found time to head up a New Haven watch company, which was one of Connecticut's largest manufacturing companies. Today, Camp is known as the "Father of American Football." He was inducted into the College Football Hall of Fame when it was founded in 1951. From 1967 until the present, the Walter Camp Player of the Year Award is given to a collegiate American football player to honor Camp's legacy.

FIRST INTERCOLLEGIATE ICE HOCKEY GAME IN THE UNITED STATES

The Yale University Bulldogs and Johns Hopkins University Blue Jays played the first United States intercollegiate ice hockey game in history in February 1896. Yale students had formed the first hockey club earlier in the 1895–96 school year, and both Yale and Johns Hopkins had been building up experience playing local athletic clubs.

Yale's top hockey players in the first game were intercollegiate tennis champions Rhode Island–born Malcolm Chace and New Haven–born Arthur Foote. Both became interested in hockey while playing in a tennis tournament in Canada. Also on that first team were New Haven native Amos Barnes, Pittsburgh's defenseman John Hall and forward William Barnett.

The Johns Hopkins team was led by team captain Canadian Sam Mitchell and Massachusetts native Rufus Bagg.

The game ended in a 2–2 tie, but a rematch two weeks later was taken by Yale 2–1. That game marked college ice hockey's first win in history. A few months later, Yale joined a hockey league and played against athletic clubs and eastern U.S. college teams. Two years later, Johns Hopkins' team folded while Yale's team continued on.

Yale player Malcolm Greene Chace (1875–1955) later became a leading financier and industrialist who was credited with bringing electrical power to the New England states. His teammate, Hillhouse High School–educated Arthur Foote (1874–1946), served as a U.S. Army officer during World War I and worked in the War Department, the American Red Cross and the Department of Commerce. He is buried in New Haven's Grove Street Cemetery.

COMMERCIAL SNOWMAKING MACHINE

Wayne M. Pierce Jr.

In the winter of 1949–50, two-year-old Mohawk Mountain ski resort in Cornwall, Connecticut, was faced with a lack of snowfall. It was a good-sized resort with nine trails, tows and parking for six hundred cars.

In late 1949, metal ski maker and former aircraft engineer Wayne M. Pierce Jr. thought he might be able to make his own snow. He enlisted the help of Arthur Hunt and Dave Richey, who were the co-owners of his Milford, Connecticut company TEY Manufacturing Corporation. (TEY Manufacturing Corporation's name was composed of the last letters of the three men's surnames.) All were also former aircraft engineers. They proceeded to make snow "from scratch" in a backyard on Ford Street in Milford with a "spray-gun nozzle, a 10-horsepower compressor and a garden hose." It copied the natural atmospheric process that creates snow.

In December 1949, the men turned their contraption loose at Mohawk Mountain. The incredible result was the first ski slopes covered with man-made snow. In December 1950, Larchmont Farms of Lexington, Massachusetts, contacted TEY Manufacturing Corporation to suggest that one of their irrigation nozzles might work well with the machine described on the TEY patent application. In 1952, the world's first major commercial snowmaking installation was set up by TEY in New York at

Grossinger's Catskill Resort Hotel. It created three inches of man-made snow each night.

In December 1950, Wayne Pierce of Milford, Connecticut, applied for a patent for "A Method for Making and Distributing Snow." It was awarded to him—assigned to TEY Manufacturing Corporation—on April 27, 1954. TEY sold the patent to Larchmont Farms, which later lost it to a Canadian researcher who had earlier unintentionally made snow while studying the effects of ice on jet engines. He hadn't produced a snowmaking machine, but he had written about his discovery in professional journals.

Today, more than 60 percent of the world's ski slopes use snowmaking equipment, while 95 percent of Mohawk Mountain's snow is man-made. It takes a lot of water to feed the machines: about 230,000 gallons of water are needed to lay a foot of snow over one acre of land.

BASEBALL CHEST PROTECTOR AND PAYPHONE INVENTOR

William Gray

Hartford, Connecticut's William Gray, who also invented the payphone, designed a baseball chest protector in 1884. By trade, he was a polisher in Hartford at Colt's armory and later at Pratt and Whitney. In his patent application to the U.S. Patent and Trademark Office, Gray described his chest protector as "an air cushion of special form, made of any air-tight material, as India-rubber cloth, the whole having transverse corrugations." He specified that it was designed for people engaged in athletic sports but that it could just as easily be adapted for use in other occupations. At first, most players wore Gray's protector under their shirts.

While Charlie Bennett was the first major-league catcher to wear a chest protector, Gray's design was the first patented, and it is the most similar design to what players use today, according to the Society for American Baseball Research.

About five years after William Gray received his chest protector patent, he invented the payphone. Co-developed by George A. Long, the first payphone was installed at the Hartford Trust Company building in downtown Hartford. At the time, it was a boon for people who could not afford their own phones—which was most people—as well as people who needed to make telephone calls away from their homes and workplaces. For

Boston Red Sox catcher George Bischoff is wearing a chest protector in this 1925 photograph. *Library of Congress.*

over one hundred years, millions of payphones could be found throughout the country (and the world). They were primarily placed in busy locations: near hospitals, train and bus stations, recreational areas like movie theaters and popular restaurants. At first, payphones cost a dime, then a quarter—only coins accepted—but later payphones took credit and debit cards. In the twenty-first century, with the advent of affordable cellphones, payphones have become almost extinct.

Chapter 9

LAW AND GOVERNMENT

A s the first state to adopt its own constitution, as well as the fifth state to approve the United States Constitution, Connecticut, as might be expected, has played a major role in the creation and development of important laws and regulations.

In this chapter, we learn about its 1600s livestock branding law, the first law of its kind in what would become the United States. Also mentioned is a significant compromise at the founding of this country—proposed by two Connecticut leaders—that made passing the U.S. Constitution possible.

Connecticut is proud of having the first law school in the country, located in Litchfield. It also was the first state to pass traffic speed limits.

FIRST BRANDING LAW

In the United States, the branding of cattle is traditionally associated with the western states. However, the first occurrence of a law requiring a brand or mark on livestock in the North American British colonies was Connecticut's branding law in 1644. Under that legislative act, all cattle, sheep and swine over six months old were required to receive ear marks to identify their owners. Since livestock was often allowed to roam about freely, this law was needed to prevent disputes among colonists on the ownership of animals.

The 1644 law stated:

> *For the preventing of differences that may arise in the owning of Cattle that be lost or stray away, It is ordered by this Court, that the owners of any Cattle within this Jurisdiction shall ear mark or brand all their Cattle and Swine that are above half a year old (except Horses,) and that they cause their several marks to be registered in the Town Book, and whatsoever Cattle shall be found unmarked after the first of July next, shall forfeit five shillings a head, whereof two shillings sixpence to him that discovers it, and the other to the Country.*

CONNECTICUT COMPROMISE

Roger Sherman and Oliver Ellsworth

In 1787, delegates from the U.S. states gathered in Philadelphia, Pennsylvania, to create a constitution for the United States. It was to replace the unworkable Articles of Confederation, which had served as the new American nation's first national constitution. For this new constitution, the convention had already developed a bicameral legislature consisting of a House of Representatives and a Senate.

However, there was one issue that came within one vote of crushing the new Constitution: the representation of the small and large states in the national Congress. The large states like New York and Virginia felt that they should have more senators in the national government than smaller states, like Rhode Island or Connecticut. Small states believed such an arrangement would mean they would be always outvoted on all issues by the large states. They wanted each state to have the same number of representatives in both the House of Representatives and the Senate.

Two ingenious members of the Constitutional Convention developed what has been called the Great Compromise. It has also been called the Connecticut Compromise since these two men were delegates from Connecticut: Roger Sherman and Oliver Ellsworth. Their solution was to have states represented in the House of Representatives by their population, while each state, no matter how large, would have two seats in the Senate. It was adopted, and from the Constitution's activation in 1789 until the present day, it has served the nation well.

Roger Sherman is also remembered as the only person to sign all of the major founding documents of the United States: the Articles of Association, the Declaration of Independence, the Articles of Confederation and the Constitution. Oliver Ellsworth later became the third chief justice of the United States Supreme Court.

FIRST SPEED LIMIT LAW IN THE UNITED STATES

Robert J. Woodruff

In 1901, Orange, Connecticut state representative Robert J. Woodruff introduced a bill to regulate the speed of automobiles in the state. The proposed law would set the speed limit of vehicles at eight miles per hour within city limits and twelve miles per hour elsewhere. After increasing the city speed limit to twelve miles per hour and the limit outside city limits to fifteen miles per hour, the Connecticut General Assembly passed "An Act Regulating the Speed of Motor Vehicles." It was the first speed limit law in the United States. The fine imposed was limited to $200 for each offense.

The law also required drivers to reduce their speed when approaching intersections or when passing vehicles drawn by horses. If the horse appeared to be "frightened," it specified that the driver of the motor vehicle must come to a stop.

Robert Woodruff only served two years in the Connecticut General Assembly. He was a graduate of New Haven's high school and earned his bachelor's and law degrees from Yale University in 1896 and 1899, respectively. Later, Woodruff became a prosecuting attorney for New Haven County and Orange's tax collector. He died in West Haven in 1958 at age eighty-four and is buried in Orange.

In 1974, due to rising fuel prices, the U.S. Congress set a national speed limit of fifty-five miles per hour for all states. As a result, the country's traffic fatality rate dropped from 4.28 per million miles traveled in 1972 to 2.73 in 1983. With fuel prices dropping back down in the 1980s, the national maximum speed limit on interstate highways was increased to sixty-five miles per hour. In 1995, speed limit control was returned to the individual states, with many of them increasing the maximum to at least sixty-five miles per hour on designated highways.

LITCHFIELD LAW SCHOOL

Tapping Reeve

Lawyer Tapping Reeve founded the first law school in America in Litchfield, Connecticut, in 1784. Prior to that, lawyers were educated through apprenticeships with practicing attorneys. However, Reeve believed that the study of the law should be separate from an undergraduate education. Before founding his law school, Reeves was best known as being part of the legal team that helped a Massachusetts enslaved woman gain her freedom by proving that the Massachusetts state constitution included the abolition of slavery.

For the first fourteen years of his school's existence, Reeves was the only instructor. He developed an eighteen-month course of formal lectures that introduced students to general legal principles. Later, he hired one of his former students, Branford, Connecticut–born James Gould, to share teaching duties. To accommodate his growing student body, Reeves built a small building adjacent to his house in Litchfield.

Litchfield Law School in Litchfield, Connecticut, which was founded by Tapping Reeve.
Collection of the Litchfield Historical Society, Litchfield, Connecticut.

Tapping Reeve's school did not grant degrees; each student received a certificate of attendance. Students were required to attend for at least three months of the fourteen-month-long program. Usually, after students graduated, they would work in a lawyer's office before being admitted to the bar. The school operated for only about half a century, but its graduates rivaled that of any law school in the history of the United States.

A list of Litchfield Law School graduates is like a who's who of early American leaders. They were highly educated, successful individuals who dominated in the public arena—especially in powerful positions in government, education and business. Among the more than one thousand students who attended Tapping Reeve's law school were future U.S. vice presidents Aaron Burr and John C. Calhoun, along with about one hundred congressmen, twenty-eight U.S. senators, six presidential cabinet secretaries, three United States Supreme Court justices and many state governors, state supreme court justices, business leaders and university presidents.

Reeve's full-time position at Litchfield Law School ended in 1798, when he was appointed to serve as a judge on Connecticut Superior Court (1798–1814) and chief judge of the Connecticut Supreme Court of Errors (1814–16).

Reeve retired from teaching in 1820 and passed away in 1823. Professor Gould continued to instruct Tapping Reeve Law School students, but plummeting enrollments forced the school to close in 1833. In its final year, there were only six students—quite a decline, considering in 1813, enrollment had reached fifty-five students. Competition from newly established university-affiliated law schools such as Harvard and Yale, which were founded in 1817 and 1826, respectively, proved to be too much for the tiny school, which no longer had Tapping Reeve at the helm.

Both Tapping Reeve's home and its law school building are still standing in Litchfield and are currently owned and managed by the Litchfield Historical Society.

FIRST FREE CHILDREN'S LIBRARY

In 1803, the first free library for children in the United States opened in Salisbury, Connecticut. The Bingham Library for Youth was begun with a gift of 150 books by Salisbury native Caleb Bingham (1757–1817), who was a noted teacher, textbook author and bookseller. At the time, Bingham explained his reason for the gift: when he was growing up in Salisbury in the 1760s, he wanted to read more, but he didn't have access to a library.

Today, on the outside wall of the Scoville Memorial Library on Main Street in Salisbury is a marker that is titled "Nation's First Public Library." Its text reads:

> *In 1803, Caleb Bingham established in Salisbury the first library in the United States open to the public free of charge. The collection was expressly created for use by young people nine to sixteen years of age although it was used by adults as well. In 1810, the town board voted to allocate tax money toward the operation of the "Bingham Library For Youth" thereby making Salisbury the first community in America to provide tax-supported public library services.*

Chapter 10

MILITARY

O ver the centuries, first as a British colony and later as one of the states of the United States, Connecticut has been at the forefront in the protection of its citizens and, after the adoption of the U.S. Constitution, the defense of individual freedoms.

The number of residents serving in the military has rivaled other states. It was small, but it was often hard to tell that by the number of volunteers and the number of soldiers and sailors who have distinguished themselves in the service of their country.

Inventions by Connecticut people have played an important part in all conflicts and wars the country has been involved in. In this chapter, we will discuss a few: the man who invented the Spencer repeating rifle, a weapon that at the time of its introduction was state-of-the art; and Colt's revolvers, which played such an important role in all wars since the nineteenth century.

On a larger scale, the first submarine ever to be used in a military action was invented by a resident of Saybrook, Connecticut, and tested several times in action during the American Revolutionary War. A century and a half later, Connecticut people in Groton built for the U.S. Navy its successor— the world's first nuclear-powered submarine.

REVOLVERS

Samuel Colt

Son of a Connecticut textile manufacturer, Hartford-born teenager Samuel Colt designed a wooden model of a revolver—that is, a gun with a revolving multichambered cylinder mechanism that could hold multiple bullets. When he was twenty-one years old, he patented a working model of the invention in England and France. The following year, he was awarded a patent in the United States. It allowed Colt to become wealthy as the sole manufacturer of revolvers until the patent expired in the 1850s.

The revolving cylinder system could be used for pistols, rifles or shotguns. In 1855, Colt built the "world's largest private armory" in Hartford, Connecticut. It was there that he and his chief engineer, Elisha Root, used a system of interchangeable parts to mass produce guns on a scale unmatched in the world. They created molds that were used to forge the individual metal pieces of the guns.

Colt's Armory in Hartford, Connecticut. This photograph was taken between 1900 and 1906. *Library of Congress.*

Colt's company produced the most widely used pistols during the Civil War. Its Colt 45 Peacemaker was the most famous pistol of the post–Civil War Wild West, acquiring the nickname "the gun that won the West." Samuel Colt died in Hartford in 1862.

In the late nineteenth century, Colt's company and its competitor, Smith and Wesson, would invent revolver cylinders that swing out from the side of the guns, allowing faster reloading of bullets. Colt Industries sold Colt Firearms Division in 1989, and it became Colt's Manufacturing Company, which is today a major manufacturer of military weapons.

Repeating Rifle

Christopher Miner Spencer

The year before the American Civil War broke out, Manchester, Connecticut–born Christopher Spencer was awarded a patent for a repeating rifle. It was the first rifle that allowed ammunition to be loaded into the rear of the gun instead of the muzzle. He sold the patent and saw a company formed to manufacture what was nicknamed the "seven shooter." He stayed on as an executive in the company. In August 1862, he took the rifle to Washington, D.C., to show it to President Abraham Lincoln. He demonstrated it for Lincoln (who personally tested it) on the "grounds below the White House" using a target board measuring six inches high and three feet wide. Impressed with Spencer's rifle, Lincoln directed the armed services to evaluate the gun, and it was used extensively by the Union army cavalry.

In 1863, Spencer furnished an infantry brigade of General William Rosecrans with his rifles. He was also a guest of General Ulysses Grant and Commodore Andrew Foote on the commodore's flagship above Vicksburg two days before the Battle of Vicksburg. After the Navy Department tried out 1,000 of the rifles, the War Department ordered 10,000. In total, the company made about 200,000 rifles. It was purchased after the war by Winchester Repeating Arms of New Haven. It was said that the company's lack of success after the war was due to the fact that so many were manufactured during the war that the post–Civil War market was flooded with used rifles.

Growing up in Manchester, Spencer was always interested in machinery, and at age fourteen, he went to work for a silk mill. That was followed by time as a journeyman machinist, machine shop superintendent and his first invention of the forty-two patents that he would be awarded during his lifetime. His first patent was for a machine that automated "the turning of spindles and bobbin heads used in sewing machine construction." That led directly to a future major invention: a "fully" automatic turret machine that could turn a small piece of metal into almost any shape without human intervention. In 1876, the Hartford Machine Screw Co. was established after Spencer invented the automatic screw machine. It quickly became one of Hartford's largest manufacturers. Later, he worked at developing a steam automobile and is said to be the first person in Connecticut to build and operate an automobile.

In Manchester in 1862, he placed a two-cylinder upright steam engine on a wagon he had designed. It made many trips between Manchester and Hartford. In about 1886, he attached a gasoline engine to a boat, making it the first powerboat to be used on the Connecticut River.

At about age eighty, Spencer invented an improved automatic screw machine and began making it for the New Britain Machine Co.

Still adventurous in his late eighties, Spencer became engrossed with aviation, making about twenty flights in the Hartford area. After spending most of his life as a resident of a home on Orchard Street in Windsor, Spencer died in Hartford at age eighty-eight.

First Military Submarine

David Bushnell

A submarine is usually defined as any naval vessel that can propel itself both beneath the water and on the water's surface. The first American submarine was invented by David Bushnell, who was born in the Westbrook section of Saybrook, Connecticut, in 1740.

Bushnell designed his first submarine in the early 1770s while he was finishing up his studies at Yale College in New Haven. He called his one-man vessel the *Turtle* because he thought it looked like two tortoise shells joined together. It included a hand-cranked screw-like propeller that moved the vessel forward and backward under water. Since gasoline-powered engines and electric motors had not been invented yet, the *Turtle* ran on man power.

When the operator wanted to descend, he would open a valve in the bottom of the submarine, and water would be let in to the ballast tanks. When he wanted to surface, he had two pumps that emptied water from the vessel. Air pipes brought fresh air into the vessel. Bushnell also demonstrated that gunpowder could explode under water and equipped the *Turtle* with a "time bomb," which would let the vessel move to a safe position before its charge went off. When the submarine reached its target, the operator was to screw an explosive into the enemy ship's hull.

In September 1776, during the American Revolutionary War, the *Turtle* was used against a sixty-four-gun British warship that was anchored in New York Harbor to assist in a blockade of the city. Its pilot, Ezra Lee, was enlisted at the last moment when its trained pilot took ill. He successfully steered the *Turtle* out to the British ship, but when he tried to attach the explosives to the side of the wooden ship, he couldn't. It is thought that this was due to either impenetrable copper sheathing on the ship or Lee's inexperience. Subsequently, he lost control of the vessel. However, the incident proved that Bushnell's invention was feasible and was an important step in the development of the submarine—for both military and civilian use.

A few years after the war—in July 1785—Thomas Jefferson, who was U.S. minister to France, wrote to Yale College president Ezra Stiles that a man in Paris, France, had invented a "method of moving a vessel on the water by a machine worked within the vessel" by using a "thin plate with its edge applied spirally round an axis." Jefferson then asked about David Bushnell:

> *The screw I think would be more effectual if placed below the surface of the water. I very much suspect that a countryman of ours, Mr. Bushnel [sic] of Connecticut is entitled to the merit of a prior discovery of this use of the screw. I remember to have heard of his submarine navigation during the war, and from what Colo. Humphreys* [David Humphreys had been Benjamin Franklin's secretary in Paris] *now tells me I conjecture that the screw was the power he used. He joined to this a machine for exploding under water at a given moment. If it were not too great a liberty for a stranger to take I would ask from him a narration of his actual experiments, with or without a communication of his principle as he should chose.*

Two months later, in response to a similar inquiry, George Washington wrote to Jefferson:

> *Bushnel [sic] is a Man of great Mechanical powers, fertile of invention, and a master in execution. He came to me in 1776 recommended by Governor Trumbull (now dead) and other respectable characters who*

were proselytes to his plan. Although I wanted faith myself, I furnished him with money, and other aids to carry it into execution....He never did succeed....I then thought, and still think, that it was an effort of genius; but that a combination of too many things were requisite, to expect much success from the enterprise against an enemy, who are always upon guard. That he had a Machine which was so contrived as to carry a man under water at any depth he chose, and for a considerable time and distance, with an apparatus charged with Powder which he could fasten to a ships bottom or side and give fire to in any given time (sufficient for him to retire) by means whereof a ship could be blown up, or sunk, are facts which I believe admit of little doubt.

Washington then gave his opinion of the problem associated with implementing Bushnell's submarine:

But then, where it was to operate against an enemy, it is no easy matter to get a person hardy enough to encounter the variety of dangers to which he must be exposed, 1. from the novelty, 2. from the difficulty of conducting the Machine, and governing it under Water on account of the Currents & ca. 3. the consequent uncertainty of hitting the object of destination, without rising frequently above water for fresh observation, which, when near the Vessel, would expose the Adventurer to a discovery, and almost to certain death. To these causes I always ascribed the non-performance of his plan, as he wanted nothing that I could furnish to secure the success of it.

Bushnell tried to use his submarine again after the Brooklyn incident, but to no effect. A few years later, he was captured by Tory forces, but when the enemy failed to recognize who he was and his importance to the American cause, he was exchanged for British soldiers. Afterward, General Washington assigned Bushnell to a top position with the U.S. Army Corps of Engineers at West Point. During the rest of the war, Bushnell worked to develop mines that could be used against British vessels.

In 1787, Bushnell left Connecticut and made his home in France, where he disappeared. Many people thought he was a victim of the French Revolution. That was, until thirty-seven years later, when a medical doctor in the state of Georgia died. It was revealed that David Bushnell had returned to the United Sates, taken the name David Bush and set up a decades-long medical practice in Georgia. The reasons behind his strange behavior have never been satisfactorily explained.

The USS *Nautilus* (SSN-571) was built by Electric Boat in Groton, Connecticut. It was the world's first nuclear-powered submarine. *U.S. Navy.*

After Bushnell's 1776 submarine, the next significant use of a U.S. submarine in warfare wasn't until the American Civil War—eighty-eight years later. In the twentieth century, during the two world wars, submarines were of great importance to the American forces. In each war, the U.S. Navy named a submarine tender after Bushnell.

FIRST NUCLEAR-POWERED SUBMARINE

U. S. Navy

USS *Nautilus* (SSN-571) was made by Electric Boat in Groton, Connecticut. Launched on January 21, 1954, it was the world's first nuclear-powered submarine. It was named after Jules Verne's fictional submarine in his 1869 classic *Twenty Thousand Leagues Under the Sea*, which, in turn, was named after the first practical submarine that was built by Robert Fulton in 1800. It was "powered by propulsion turbines that were driven by steam produced by a nuclear reactor." The submarine's construction took only one year, seven months and seven days.

The USS *Nautilus* was christened on January 21, 1954, by First Lady Mamie Eisenhower as thirty thousand attended the ceremony at the Electric Boat Division of General Dynamics Corporation in Groton. She was the first U.S. president's wife to ever christen a navy submarine. The official program for the launching promised that the *Nautilus* "will cruise submerged faster, farther, longer than any previous craft in history." It was also larger than the diesel-electric submarines used during World War II.

Eight months later, it was commissioned, and Commander Eugene P. Wilkinson was its first commander. The specifications of the ship were:

> *323 feet long (98 meters)*
> *Displaced 4,092 tons when submerged*
> *Crew of 104*
> *Top speed on the surface 22 knots*
> *Top speed underwater 20+ knots*

In 1955, after sea trials, *Nautilus* headed south to Puerto Rico. It traveled 1,381 miles while submerged in 89.8 hours. It was the longest submerged cruise ever made by a submarine. The *Nautilus* also had the highest "sustained submerged speed ever recorded for a period of more than one hour's duration."

In another historic cruise, the *Nautilus*, on August 1–5, 1958, passed beneath the polar ice cap in a 1,830-mile-long voyage from Point Barrow, Alaska, to the Greenland Sea.

The *Nautilus* was decommissioned in 1980 and in 1985 went on exhibit at the USS Nautilus Memorial and Submarine Force Library and Museum in Groton, Connecticut.

Chapter 11

INNOVATIONS

In this chapter, we will discuss Connecticut people who have played important roles in helping other people. Included is the man who is often spoken of as the founder of the modern American consumer movement, as well as another innovator who founded a nonprofit organization to lend assistance to anyone in the world who is the victim of a natural disaster.

We begin with a man who had great influence on the education of all Americans: the creator of the first American dictionary.

AMERICAN DICTIONARY

Noah Webster

Noah Webster was born in 1758 in the western part of Hartford, which would become the town of West Hartford almost a century later. After graduating from Yale College in 1778, he became a teacher and studied law. He was admitted to the bar in 1781. In New Haven, Webster lived at the intersection of Temple and Grove Streets. A marker is embedded in the outside wall of a dormitory of Yale University's Silliman College that reads:

Noah Webster was the creator of the *American Dictionary of the English Language*. *National Portrait Gallery, Smithsonian Institution.*

Here stood the house of
Noah Webster
Class of 1778
Author of The American
Spelling Book and of An American
Dictionary of the English Language

Webster worked at recognizing that the American English language and the American educational systems were uniquely different from their British equivalents. First among his endeavors was to write *The American Spelling Book* in 1783, which by the end of the century had sold 100 million copies. Multiple generations used it to learn to read and spell; it also had lessons on the American government and morality.

Chief among Webster's ventures was the formulation of the first American dictionary in 1806. This was continually improved until, by 1828, his *American Dictionary of the English Language* was over sixty-five thousand words long. Webster used many American spellings that were different than the English dictionaries,

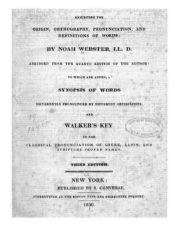

The title page of 1830's *American Dictionary of the English Language* by Noah Webster. *Library of Congress.*

such as *color* instead of *colour*, and added words that were common in the American vocabulary that were not found in the English dictionaries, such as *census, checkers, debit, hickory, immigrant, psychology, publicity, skunk, slang, vaccine* and *whiskey*. Webster also included words from law, medicine, the natural sciences and Native American languages that were used in the United States but were not included in other dictionaries.

After Webster's death in New Haven at age eighty-four, George and Charles Merriam purchased rights to his dictionary from his estate. The Merriam brothers hired Noah's son-in-law Chauncey Goodrich to be editor-in-chief and Noah's son William to be editor of the first Merriam-Webster dictionary in 1847.

OLDEST CONTINUOUSLY PUBLISHED NEWSPAPER IN THE UNITED STATES

Thomas Green

From 1760 through 1764, New London–born Thomas Green was editor of New Haven's first newspaper, the *Connecticut Gazette* (begun in 1755), as well as serving as the city's postmaster.

In the late summer of 1764, Green moved to Hartford to start his own business. He set up an office on Main Street in a building over a barbershop and later moved to another well-frequented location next to the courthouse and Bull's Tavern., The weekly newspaper he founded was called the *Connecticut Courant*. Today, it is known as the *Hartford Courant* and is the oldest continuously published paper in the United States.

Each Saturday, letters, newspapers and flyers arrived at Green's Hartford office from New York City and Boston. Like many papers of eighteenth-century Connecticut, overseas news occupied much of the paper, while relatively little space was given to local news. In addition to producing the *Connecticut Courant*, Green printed broadsides and pamphlets and diversified into the sale of books (including Bibles and spelling books), along with stationery items.

Green only stayed in Hartford for a few years before turning over the ownership of his paper to others and returning to New Haven. Many years later, the main conference room at the *Hartford Courant* was named the Thomas Green Room.

Miss Porter's School

Sarah Porter

In 1843, Farmington's thirty-year-old Sarah Porter founded Miss Porter's School in her hometown. It became one of the most respected preparatory schools for girls in the country.

The daughter of a Congregational Church minister, Porter was educated at Farmington Academy, where she was the only female student, and later was tutored by Yale College professors. At Miss Porter's School, she taught a wide range of subjects. In later years, the school's courses included Latin, French, German, history, geography, music,

Sarah Porter founded Miss Porter's School, a preparatory school for girls, in Farmington, Connecticut. *National Portrait Gallery, Smithsonian Institution.*

philosophy, algebra, trigonometry, chemistry, botany, geology and astronomy. Bible studies took place on Sundays.

Sarah Porter continued to teach at her school until she was in her eighties. She died in Farmington in 1900.

Many of the most prominent and accomplished women in the United States were graduates of Miss Porter's School for Girls, including First Lady Jacqueline Bouvier Kennedy, who attended from 1944 to 1947.

KNIGHTS OF COLUMBUS

Michael J. McGivney

In 1882, thirty-year-old Catholic priest Michael J. McGivney and a group of his parishioners at St. Mary's Church in New Haven, Connecticut, founded the Knights of Columbus. The Waterbury-born McGivney had been ordained a priest only five years earlier. Its main objective was to provide assistance to the organization's members and their families who were sick, disabled or otherwise in need through insurance and other means. It also provided social and volunteer activities for its members.

Statue of Knights of Columbus founder Reverend Michael J. McGivney in front of St. Mary's Catholic Church in New Haven, Connecticut. *Author's collection.*

Today, the organization is the largest Roman Catholic fraternal benefit society in the world. The world headquarters of the society's supreme council is in New Haven, about one mile from St. Mary's Church. The latter is the final resting place of Father McGivney, who in 2020 was beatified (the last step before canonization) by Pope Francis.

In recent years, the Knights of Columbus has partnered with such charities as Habitat for Humanity and the Special Olympics to provide help to people in need of all faiths and cultures.

LANDSCAPE ARCHITECTURE

Frederick Law Olmsted Sr.

Known today as the "Founder of American Landscape Architecture," Frederick Law Olmsted Sr. was born in Hartford, Connecticut, in 1822. Most of his grammar school education was obtained from ministers in towns near Hartford. Most memorable for him were the trips he took with his family through scenic areas of northern New England and New York State. Olmsted was to write years later:

> *The happiest recollections of my early life are the walks and rides I had with my father and the drives with my father and stepmother in the woods and fields. Sometimes these were quite extended, and really tours in search of the picturesque. Thus, before I was twelve years old I had been driven over the most charming roads of the Connecticut Valley and its confluents, through the White Hills and along most of the New England coast from the Kennebeck to the Naugatuck.*

Olmsted's connections with Connecticut are many—attending school in Collinsville (part of the town of Canton), staying with relatives in Cheshire and trying experimental farming in Guilford.

As he grew older, Olmsted engaged in diverse experiences that would serve him well for his future careers: clerking in a dry goods store, attending lectures at Yale University, managing a farm on Long Island, taking a year-long voyage to China and extensively walking through Europe and the British Isles (where he observed landscapes and public and private parks). He wrote about the latter in his book about his observations, *Walks and Talks of an American Farmer in England* (1852).

A firm opponent of slavery, Olmsted traveled as a newspaper reporter in America's southern states in the early 1850s and sent back weekly reports on their economy and the position of the enslaved people. In 1857, he obtained the position of superintendent of Central Park on Manhattan Island. As such, he was in charge of the park's design and construction projects. This was followed by his service as director of the U.S. Sanitary Commission, which had authority over the "health and camp sanitation" of all the volunteer soldiers of the Civil War's Union army.

In private practice in the landscaping business, Olmsted and his associates worked with engineers, horticulturists and other specialists, as well as undertaking small, midsized and large projects throughout the United States. Olmsted's landscape architecture not only would forever change America's landscape, but it would also impact the American people's quality of life forever. The Olmsted projects included government property, such as the U.S. Capitol grounds and terraces in Washington, D.C.; and public and private parks, college campuses and private estates (such as the huge Biltmore Estate in Asheville, North Carolina). In Connecticut, projects included Walnut Hill Park in New Britain (1870), the Connecticut State House in Hartford (1878), Seaside and Beardsley Parks in Bridgeport (1884), the property of what is now the Institute for Living in Hartford and the grounds of the Wadsworth Mansion in Middletown.

Olmsted retired in 1895 and passed away in 1903. He is buried in Hartford. In 2021, the State of Connecticut's Office of Economic and Community Development presented the private nonprofit organization Preservation Connecticut with a grant to do a survey of Connecticut's Frederick Law Olmsted's landscape legacy. This was to mark the 200th anniversary of Frederick Law Olmsted's birth in 2022.

SELF-HELP BOOKS

P.T. Barnum

The number of self-help titles that have been published is fast approaching 100,000. Each year, billions of dollars are spent in the United States alone on books that promise the solutions to personal problems. They will teach you how to be a leader, how to improve your relationships, how to maintain a healthy lifestyle and so on. But who published the first self-help book, and what was its subject?

Many people credit nineteenth-century showman P.T. Barnum with inventing the self-help industry with his 1880 book *The Art of Money Getting*. The book's chapter titles include "Avoid Debt," "Do Not Scatter Your Powers," "Be Systematic," "Advertise Your Business" and "Be Polite and Kind to Your Customers."

Barnum is of course known for founding, with James A. Bailey, the most famous circus in history: Barnum & Bailey Circus, which was ultimately known as the Ringling Bros. and Barnum & Bailey Circus. Barnum created it as a huge spectacle and dubbed it the "Greatest Show on Earth."

But in addition to his circus, Barnum developed the public museum, the musical concert and one of the most

This statue of showman P.T. Barnum was erected at the Bethel Public Library on the 200th anniversary of Barnum's birth. *Author's collection.*

popular books about how to succeed. An example of the advice in the book is: "But let money work for you, and you have the most devoted servant in the world. It is no 'eye-servant.' There is nothing animate or inanimate that will work so faithfully as money when placed at interest, well secured. It works night and day, and in wet or dry weather."

THE FIRST GRADUATE SCHOOL DEDICATED TO FORESTRY IN AMERICA

In 1900, at New Haven's Yale University, Henry S. Graves (1871–1951), James Toumey (1865–1932) and Gifford Pinchot (1865–1946) founded the first graduate school dedicated to forestry in America.

Henry Graves became the school's first director and first professor, serving until 1910. He had already earned both a bachelor's (1896) and a master's (1900) degree from Yale. After his stint as the forestry school's director, Graves became the second chief of the United States Forest Service. After ten years in that position, he returned to New Haven to become dean of the forestry school for sixteen years (1923–39).

The second founder, James Toumey, took over for Henry Graves as acting dean (and later dean) of Yale's forestry school when the latter became chief of the United States Forest Service in 1910. Twelve years later, Graves returned to his old position, but Toumey stayed with the forestry school until his death in 1932.

The third founder, Simsbury, Connecticut native Gifford Pinchot, became the first chief of the U.S. Forest Service in 1905 and served in that role until 1910 under U.S. presidents Roosevelt and Taft. After graduating from Yale University in 1889, he became a resident forester for George Washington Vanderbilt's Biltmore estate and worked at the National Forest Commission. In later life, Pinchot served two nonconsecutive terms as the governor of Pennsylvania.

"Father of U.S. Consumer Protection"

Ralph Nader

For over fifty years, Ralph Nader's name has been synonymous with consumer protection. Born in Winsted, Connecticut, to Lebanese immigrants, he received his bachelor's degree from Princeton University and graduated from Harvard Law School. After several years as a lawyer in Hartford, Connecticut, Nader became interested in the issue of automobile safety.

At age thirty, Nader worked as a consultant to the U.S. Department of Labor. The following year, his book *Unsafe at Any Speed* was published. One of the most influential books of twentieth-century America, it revealed serious safety concerns regarding U.S. automotive industry vehicles. The book inspired 1966's National Traffic and Motor Vehicle Safety Act, which allowed the federal government to mandate safety measures for all motor vehicle manufacturers.

Nader once said, "This country has far more problems than it deserves and far more solutions than it applies." Over the years, Nader attracted consumer activists, popularly known as Nader's Raiders, who assisted him on issues ranging from the regulation of insecticides to the safety of nuclear plants, from meat processing to international trade. The effectiveness of their research and lobbying efforts is shown by the many changes they saw enacted into federal, state and local laws and regulations.

In his later years, Ralph Nader published numerous books on consumer advocacy and other political topics from his home in Winsted, Connecticut.

Legendary consumer advocate Ralph Nader speaking in New York City in 2013. *Author's collection.*

In 2014, he founded the American Museum of Tort Law in Winsted as the first law museum in the United States. The institution's stated mission is very similar to Nader's life mission: "To educate, inform and inspire Americans about two things: Trial by jury; and the benefits of tort law. Tort law is the law of wrongful injuries, including motor vehicle crashes, defective products, medical malpractice, and environmental disasters, among many others."

AMERICARES

Robert Macauley

Robert Macauley founded the immediate response disaster relief and humanitarian aid organization Americares in Stamford, Connecticut, in 1979. Concentrating on health, it has provided relief to millions of people harmed by poverty and disasters. Since its founding, it has given more than $20 billion in humanitarian aid to the needy both inside and outside the United States. It provides free clinics in Connecticut, which for almost thirty years have furnished help to its people.

Born in New York City, Macauley grew up in Greenwich, Connecticut, attending Greenwich Country Day School, and began studies at Yale University in New Haven. When World War II broke out, he served in the U.S. Army Air Corps. After the war's end, he returned to Yale, where future president George H.W. Bush was his roommate. After graduation from Yale, Macauley served as an executive in the paper industry.

Americares' story began in April 1975, when the United States was preparing to leave South Vietnam. A U.S. jet with more than two hundred orphans aboard crashed into the jungle. The U.S. military declared that it would not be able to rescue them for eleven days. Robert Macauley, who in the past had donated to a charity to help orphans of the Vietnam War, personally stepped in, taking out a $10,000 mortgage on his house in order to lease a Boeing 747 jet to airlift a total of three hundred orphans to the United States.

Six years later, Macauley was asked by Pope John Paul II to deliver medical supplies to Poland, which was under martial law. Macauley, using his extensive connections, convinced companies to donate medical supplies and had thirty-eight airlifts of supplies delivered to the people of Poland by the following March.

In 1982, Robert Macauley founded Americares and became its CEO—a position he held for twenty years without accepting a salary. He was chairman of the board of Americares until his death in 2010 at age eighty-seven. At that time, he had a home in New Canaan, Connecticut. During his time leading Americares, Macauley always pushed to respond to events as soon as possible. He believed that the sooner aid arrived, the more lives would be saved and the less suffering would be incurred. Just a few of the major disasters that Americares has been at the forefront of providing aid for are:

- The Ethiopian famine of 1985
- The Armenian earthquake of 1988
- Kuwait after it was invaded by Iraq in 1990
- The aftermath of the attack on New York City's World Trade Center in 2001
- The Haiti earthquake of 2010
- Joplin, Missouri's tornado of 2011
- The COVID-19 pandemic, beginning in 2020
- The war in Ukraine in 2022

According to *Forbes* magazine, in 2021 Americares was the seventh-largest U.S. charity. That year, it received nearly $1.5 billion in private donations and only $5 million in government support.

BIBLIOGRAPHY

"An Act Regulating the Speed of Motor Vehicles." In *Public Acts Passed by the General Assembly of the State of Connecticut in the Year 1901*. Hartford, CT: Belknap & Warfield, 1901.

Administrator, Olmsted Parks. "Frederick Law Olmsted Sr." National Association for Olmsted Parks. www.olmsted.org/the-olmsted-legacy/frederick-law-olmsted-sr.

American Artifacts. "Report on Mr. L.E. Denison's Corn Sheller." www.americanartifacts.com/smma/sheller/denison.htm.

American History. "Ken Olsen Interview." americanhistory.si.edu/comphist/olsen.html.

The American Museum of Tort Law. "About Us." November 2, 2021. www.tortmuseum.org/about-us.

Americares. "Frequently Asked Questions." www.americares.org/our-faq.

ASME. "Elias Howe." www.asme.org/topics-resources/content/elias-howe.

———. "Eli Terry." www.asme.org/topics-resources/content/eli-terry.

Baker, Kendra. "Connecticut's Speed Limit Was Created 120 Years Ago and It Was the First to Exist in the U.S." *NewsTimes*, May 21, 2021. www.newstimes.com/local/article/Connecticut-s-speed-limit-was-created-120-years-16192169.php.

The Barnum Museum. "P.T. Barnum, the Man, the Myth, the Legend." barnum-museum.org/about/the-man-the-myth-the-legend.

Belleville, Michelle. "About the Hubble Space Telescope." NASA, September 24, 2019. www.nasa.gov/mission_pages/hubble/about.

Bigelow Tea. "Meet the Bigelow Family." www.bigelowtea.com/Our-Family-Story/Meet-the-Bigelows.

Boyd, Herb. "Sarah Boone, Inventor of the Ironing Board and First Black Woman to Get a Patent." *New York Amsterdam News*, October 22, 2021. amsterdamnews.com/news/2021/08/05/sarah-boone-inventor-ironing-board-and-first-black.

Bulletin of the American Astronomical Society. "Obituary: E. Dorrit Hoffleit, 1907–2007." Vol. 39, no. 4 (200): 1067–69.

CBIA. "Bloomberg Ranks Connecticut Fourth for Innovation." May 12, 2022. www.cbia.com/news/economy/bloomberg-innovation-index-ranks-connecticut-fourth.

Christian Science Monitor. "Old Time to South America Halved by New Air Schedules: News in Aviation One Day to Canal East Coast Route Routes to Interior 'Vertaplane' Gets Trial Germany Leads at Air Meet." August 3, 1937, 5.

Columbia Bicycles. "Columbia Bicycles Heritage—America's First Bicycle." April 12, 2018. columbiabicycles.com/heritage.

Connecticut Explored. "Peter Paul's Path to Sweet Success." September 29, 2020. www.ctexplored.org/peter-pauls-path-to-sweet-success.

Connecticut History: A CThumanities Project. "Bevin Brothers Helps Transform East Hampton into Belltown, USA." December 8, 2020. connecticuthistory.org/bevin-brothers-helps-transform-east-hampton-into-belltown-usa.

———. "Charles Ritchel and the Dirigible." February 22, 2022. connecticuthistory.org/charles-ritchel-and-the-dirigible.

———. "The Oldest Continuously Published Newspaper." October 25, 2021. connecticuthistory.org/the-oldest-continuously-published-newspaper-today-in-history.

———. "Reel Lawn Mower Patent." January 28, 2018. connecticuthistory.org/reel-lawn-mower-patent-today-in-history.

Doyle, Nancileigh M., et al. "The Life and Work of Harvey Cushing 1869–1939: A Pioneer of Neurosurgery." *Journal of the Intensive Care Society* 18, no. 2 (2016): 157–58. doi.org/10.1177/1751143716673076.

1885 Yale College Obituary Record. "Obituary of Alexander Catlin Twining, Class of 1820."

Eisenhower Presidential Library. www.eisenhowerlibrary.gov/sites/default/files/research/online-documents/uss-nautilus/program.pdf.

Elbaum-Garfinkle, Shana. "Close to Home: A History of Yale and Lyme Disease." *Yale Journal of Biology and Medicine* 84, no. 2 (2011): 103–8.

Encyclopædia Britannica. "Charles Goodyear." www.britannica.com/biography/Charles-Goodyear.

———. "Ralph Nader." www.britannica.com/biography/Ralph-Nader.

Eno, William Phelps. *Street Traffic Regulation; General Street Traffic Regulations—Special Street Traffic Regulations…Dedicated to the Traffic Squad of the Bureau of Street Traffic of the Police Department of the City of New York*. N.p.: Rider and Driver Publishing Co., 1909.

Eofgang@connecticutmag.com. "Connecticut's Original Electric Car." *CT Insider*, March 30, 2022. www.ctinsider.com/connecticutmagazine/news-people/article/Connecticut-s-Original-Electric-Car-17042927.php.

Forest History Society. "Henry S. Graves (1871–1951)." February 27, 2020. foresthistory.org/research-explore/us-forest-service-history/people/chiefs/henry-s-graves-1871-1951.

Frank J. Sprague Papers. archives.nypl.org/mss/2850.

Frank Pepe Pizzeria. "About Us." pepespizzeria.com/about.

Gregg, Helen. "His Current Quest." *University of Chicago Magazine*. mag.uchicago. edu/science-medicine/his-current-quest.

Harkins, John. "Hiram Percy Maxim." *Life*, September 1932, 12–15.

Hartford Courant. "Year's Patent Record: Connecticut Still Leads the Country." March 15, 1898, 2.

Hastings, Charles S. *Josiah Willard Gibbs: 1839–1903*. nasonline.org/publications/ biographical-memoirs/memoir-pdfs/gibbs-josiah.pdf.

The Henry Ford. "Frisbie's Pies Plate, circa 1935." www.thehenryford.org/ collections-and-research/digital-collections/artifact/350954.

———. "Westinghouse Lightning Arrester, 1889–1891." www.thehenryford.org/ collections-and-research/digital-collections/artifact/201587.

Hirahara, Naomi. *Distinguished Asian American Business Leaders*. Westport, CT: Greenwood Press, 2003.

Hubbell. "About Hubbell." www.hubbell.com/hubbell/en/about-hubbell.

Internet Hall of Fame. internethalloffame.org/inductees/vint-cerf.

Ives, Frederic Eugene. *Hand-Book to the Photochromoscope, by Its Inventor…with Chapters on the Nature of Light, the Theory of Color, Etc*. London, 1894.

Jackson, Edith B. "The Development of Rooming-In at Yale." *Yale Journal of Biology and Medicine* 25 (1953): 484–94.

Jerome, Chauncey. *History of the American Clock Business for the Past Sixty Years: And Life of Chauncey Jerome*. N.p., 1860.

Knight, Margaret E. "Paper-Bag Machine." October 28, 1879.

The Lancet. "Orvan Hess." Vol. 360, no. 9340, October 12, 2002, 1179.

Litchfield Historical Society. "Tapping Reeve House and Litchfield Law School." www. litchfieldhistoricalsociety.org/museums/tapping-reeve-house-and-law-school.

Markoff, John. "Wesley A. Clark, 88, Dies; Made Computing Personal." *New York Times*, February 28, 2016, 24.

Mars. "Our History." www.mars.com/about/history.

Merriam-Webster. "About Us." www.merriam-webster.com/about-us/reform-glossary.

Morga, Adriana. "Connecticut Inventions: 20 Things That Were Created or Patented in the Nutmeg State." *CT Insider*, August 11, 2021. www.ctinsider. com/living/article/Inventions-created-in-Connecticut-16379555.php.

Museum of Connecticut History. "Connecticut Patents." May 13, 2021. museumofcthistory.org/patent/connecticut-patents.

Myers, J.S. *Life and Letters of Dr. William Beaumont*. Concord, NH: Rumford Press, 1913.

National Archives and Records Administration. "Eli Whitney's Patent for the Cotton Gin." www.archives.gov/education/lessons/cotton-gin-patent.

———. "Founders Online: From Thomas Jefferson to Ezra Stiles, 17 July 1785." founders.archives.gov/documents/Jefferson/01-08-02-0236.

———. "Founders Online: To Thomas Jefferson from David Bushnell, [13] October 1787." founders.archives.gov/documents/Jefferson/01-12-02-0292.

National Aviation Hall of Fame. "Frederick Brant Rentschler." nationalaviation. org/enshrinee/frederick-brant-rentschler.

National Institutes of Health, U.S. Department of Health and Human Services. "Gallo, Robert C.—History—Office of NIH History and Stetten Museum." history.nih.gov/pages/viewpage.action?pageId=11600169.

National Inventors Hall of Fame. "NIHF Inductee Leonard Flom Invented the Iris Recognition System." www.invent.org/inductees/leonard-flom.

———. "NIHF Inductee Leopold Godowsky and Kodachrome Film Processing." www.invent.org/inductees/leopold-godowsky-jr.

National Science Foundation. "Barbara McClintock (1902–1992)." www.nsf.gov/news/special_reports/medalofscience50/mcclintock.jsp.

Naval History and Heritage Command. "*Nautilus* (SSN-571)." www.history.navy.mil/browse-by-topic/ships/submarines/uss-nautilus.html.

New England Historical Society. "Elias Howe, Sewing Machine Inventor, Gets a Little Help from the Beatles." July 10, 2021. www.newenglandhistoricalsociety.com/elias-howe-sewing-machine-inventor-gets-little-help-beatles.

New York Times. "James Brunot, 82, the First Producer of Scrabble Games." October 27, 1984. www.nytimes.com/1984/10/27/obituaries/james-brunot-82-the-first-producer-of-scrabble-games.html.

Nicholas, J.S. *Ross Granville Harrison: 1870–1959*. www.nasonline.org/publications/biographical-memoirs/memoir-pdfs/harrison-ross.pdf.

Nobel Prize. "The Nobel Prize in Chemistry 1989." www.nobelprize.org/prizes/chemistry/1989/altman/biographical.

North, Simon Newton Dexter, and Ralph H. North. *Simeon North: First Official Pistol Maker of the United States: A Memoir*. Concord, NH: Rumford Press, 1913.

Pepperidge Farm. "Our Story." March 1, 2021. www.pepperidgefarm.com/our-story.

Pitney, A.H. "Postage Meter and Mail Marking Machine." Patent 1,370,668. March 8, 1921.

Pitney Bowes. "100 Years of Pitney Bowes." www.pitneybowes.com/us/100years.html.

Queen Elizabeth Prize for Engineering. "Digital Imaging Sensors." qeprize.org/winners/digital-imaging-sensors.

Ray, James Lincoln. "Don Mattingly." Society for American Baseball Research, January 7, 2022. sabr.org/bioproj/person/don-mattingly/.

Schiff, Judith Ann. "History on Ice: The First Intercollegiate Hockey Game." *Yale Alumni Magazine*, February 2003. archives.yalealumnimagazine.com/issues/03_02/old_yale.html.

Sikorsky Archives. "Civilian Rescue." www.sikorskyarchives.com/Civilian_Rescue.php.

Silverstein, Barry. "Why Did the World's Second Largest Computer Company Fail?" *Medium*, April 3, 2022. historyofyesterday.com/why-did-the-worlds-second-largest-computer-company-fail-cf9977152835.

Simmons, Amelia. "American Cookery: The Art of Dressing Viands, Fish, Poultry, and Vegetables." Project Gutenberg, April 6, 2022. www.gutenberg.org/cache/epub/12815/pg12815.html.

Smithsonian Institution. "Meet Mary Kies, America's First Woman to Become a Patent Holder." May 5, 2016. www.smithsonianmag.com/smart-news/meet-mary-kies-americas-first-woman-become-patent-holder-180959008.

Spencer Turbine. "Company History." May 23, 2019. www.spencerturbine.com/about-us/company-history.

State of Connecticut Judicial Branch. "Tapping Reeve and the Litchfield Law School." www.jud.ct.gov/lawlib/history/tappingreeve.htm.

Sullivan, Ronald. "William A. Higinbotham, 84; Helped Build First Atomic Bomb." *New York Times*, November 15, 1994. www.nytimes.com/1994/11/15/obituaries/william-a-higinbotham-84-helped-build-first-atomic-bomb.html.

Thoms, Herbert. "Natural Childbirth in a Teaching Clinic." *BJOG: An International Journal of Obstetrics and Gynaecology* 56, no. 1 (1949): 18–21. doi.org/10.1111/j.1471-0528.1949.tb07067.x.

University of Utah Health. "The First Artificial Heart, 30 Years Later." healthcare.utah.edu/healthfeed/postings/2012/12/120212ArtificialHeart30YearsLater.php.

U.S. Patent and Trademark Office. "General Information Concerning Patents." May 12, 2022. www.uspto.gov/patents/basics/general-information-patents.

———. "1994 Laureates: National Medal of Technology and Innovation." October 15, 2021. www.uspto.gov/learning-and-resources/ip-programs-and-awards/national-medal-technology-and-innovation/recipients/1994.

U.S. Patent Full-Text Database Number Search. patft.uspto.gov/netahtml/PTO/srchnum.htm.

U.S. Patent Office. Annual Report of the Commissioner of Patents for the Year 1897. library.si.edu/digital-library/book/annualreportofco1897unit.

Waggoner, Walter H. "James Brunot, 82, the First Producer of Scrabble Games." *New York Times*, October 27, 1984. www.nytimes.com/1984/10/27/obituaries/james-brunot-82-the-first-producer-of-scrabble-games.html.

Weller, Thomas H., and Frederick C. Robbins. *John Franklin Enders: 1897–1985*. National Academy of Sciences. www.nasonline.org/publications/biographical-memoirs/memoir-pdfs/enders-john.pdf.

Wells, Horace. *A History of the Discovery of the Application of Nitrous Oxide Gas, Ether, and Other Vapors to Surgical Operations*. Hartford, CT: J. Gaylord Wells, 1847.

Wilson, P.K. "Daniel Turner and the Art of Surgery in Early Eighteenth-Century London." *Journal of the Royal Society of Medicine* (December 1994). www.ncbi.nlm.nih.gov/pubmed/7853312.

Witkowski, Mary K. "Lewis H. Latimer, African American Inventor." Bridgeport History Center. bportlibrary.org/hc/african-american-heritage/louis-latimer.

Yale, L. "Padlock." Patent 18,169. September 8, 1857.

Yale New Haven Hospital. "Learning to Care for Your Newborn at Yale New Haven Hospital." www.ynhh.org/services/maternity-services/childbirth-resources/learning-to-care-for-your-newborn.

———. "Yale New Haven Children's Hospital Opens One of the Most Advanced Neonatal Intensive Care Units in the United States." www.ynhh.org/news/ynhch-opens-one-of-the-most-advanced-neonatal-intensive-care-units-in-the-united-states.aspx.

Yale School of Medicine. "Fulton, Penicillin and Chance." December 15, 1999. medicine.yale.edu/news/yale-medicine-magazine/fulton-penicillin-and-chance.

ABOUT THE AUTHOR

Peter Hubbard is a computer information systems instructor at Post University in Waterbury, Connecticut. His background is in Internet research, cybersecurity and web design. He was born in Connecticut and has lived in New Haven County for over thirty years.